Knowledge Production Modes between Science and Applications 1

Series Editor
Jean-Charles Pomerol

Knowledge Production Modes between Science and Applications 1

Concepts

Jean-Claude André

iSTE

WILEY

First published 2024 in Great Britain and the United States by ISTE Ltd and John Wiley & Sons, Inc.

ISTE Ltd
27-37 St George's Road
London SW19 4EU
UK

www.iste.co.uk

John Wiley & Sons, Inc.
111 River Street
Hoboken, NJ 07030
USA

www.wiley.com

Library of Congress Control Number: 2023946437

British Library Cataloguing-in-Publication Data
A CIP record for this book is available from the British Library
ISBN 978-1-78630-807-8

Contents

Foreword

Additive Manufacturing: From 3D Printing to Bio-printing

Jean-Claude André's new book is invaluable because it is written by someone who has a wealth of experience and has studied and followed the path from the researcher's idea to industrial application.

For over 50 years, Jean-Claude André has successively or simultaneously been a researcher, deputy scientific director of the CNRS engineering sciences department, scientific director of a joint organization devoted to workplace safety and advisor to the Institut national des sciences et de l'ingénierie des systèmes (INSIS-CNRS). It is fortunate that he thought to put his experience in writing because, unfortunately, in France, the invaluable experience of people like Jean-Claude André is often ignored by "managers" with no culture of innovation and often even without the slightest knowledge in science and/or engineering. May they one day open this book, promote innovation and halt the deindustrialization of France.

Jean-Claude André begins by reminding us of the role of manufacturing in a country's economy, and its dependence on science and innovation. He points out – and no one should ignore this – that a job created in an industrial zone, as a result of political incentives and subsidies, and inaugurated with much fanfare, costs the community at least 10 times more than a job created in an innovative startup.

A large part of the book (in two volumes) is devoted to the transition from invention to innovation to industrial manufacturing. Chapters 1 and 2 of Volume 1 follow the long and winding road from one to the other, from supporting creativity to mobilizing producers, via the indispensable POC (proof of concept). Jean-Claude André describes in detail all of the pitfalls to be avoided in transforming an idea into a product. We must thank him for shedding light on the decision-making process, the immoderate and sometimes ignorant use of TRLs (Technology Readiness Levels) by the research administration in France and Europe. He also devotes several pages to industrial property and patents, indispensable for private research, yet disastrous as performance indicators for researchers since they measure above all an institution's ability to pay lawyers and annuities. A great deal of space is devoted to the startups that line the road from idea to unicorn, many of whose corpses lie by the side of the road. We look at the various ways in which we can help them avoid going off the rails.

In Volume 2, Chapter 1 focuses on the social responsibility of researchers, inventors and companies. Communication plays a key role as an interface between science and the public. The difficulties and weaknesses highlighted during the Covid-19 crisis are well known. The author does not shy away from the issue, stressing the extent to which communication errors, ill-considered announcements made by ignorant people, the procrastination of expert advice and false hopes fuel the resentment of the crowd, always quick to look for scapegoats.

Beyond individual responsibility, Jean-Claude André advocates socially responsible research (SRR). At the end of the first part of this two-volume work (Chapters 1 and 2 of Volume 1 and Chapter 1 of Volume 2), the author summarizes in a table (section 1.5, Volume 2) that all of those responsible for innovation policy should have in front of them lists of the 12 obstacles to overcome in order to move from the researcher's idea to innovation and economic success.

Chapter 2 of Volume 2 looks back at Jean-Claude André's experience as inventor, with Alain Le Méhauté, of 3D printing, now known as "additive manufacturing". Jean-Claude André describes the genesis and development of his adventure, which led from a laboratory experiment to additive manufacturing and its various technologies. An adventure with sales in the billions of euros per year, which is far from over. It is a fascinating story that sheds light on the blindness of many institutions and officials who prefer to do nothing, rather than take the slightest risk of supporting an innovation that could ultimately fail. It is as if we should only support startups that we are sure will become unicorns!

From 3D printing, Jean-Claude André moves on to 4D, describing the problems and associated technologies. He explains how this type of innovation must initially rely on a niche policy. After 4D printing, we move on to bio-printing and its challenges.

Chapter 3 of Volume 2 provides an overview of all of the processes that lead from creativity to industrialization. Asking the right questions, valuing interaction between researchers, fostering interdisciplinarity and integration are just some of the recommendations that emerge, drawn from the experience of starting up 3D printing, without forgetting the role of chance. The future of 4D and bio-printing is considered in the light of the applications that are already possible and described.

All in all, this book contains the invaluable lessons of real-life experience and encyclopedic knowledge, backed up by a truly impressive bibliography. Nowhere else will you find such an important body of knowledge on the links between research, invention and industrialization. In addition, Jean-Claude André's erudition is also reflected in an impressive number of quotations, sometimes surprising or amusing, but always highly illuminating, making this book an enthralling read, dense but exceptional in its truth of a lifetime's experience.

Jean-Charles POMEROL
President of the Agoranov incubator

Preface

There is no doubt, Gentlemen, that the public, as a whole, is entitled to the regular and proven products of literary industry, but the advancement of industry requires many attempts, daring hypotheses, imprudence even; and laboratories alone make it possible to achieve the very high temperatures, the rare reactions, the degrees of enthusiasm without which neither the sciences nor the arts would have anything but an over-anticipated future.

(Valéry 1957)

Science marches on, making unprecedented progress every day. Every day our knowledge expands in depth and surface, and at the same time the moral ideal grows. In vain will we struggle against the evidence; it is science that leads the world. Humanity will know no other guide.

(Richet 1895)

Why is philosophy important to engineering? Ultimately and most deeply it is because engineering is philosophy – and through philosophy engineering will become more itself.

(Mitcham 1998)

Technological progress induces a dynamic of movement, insofar as it possesses a directional quality that makes the state of arrival preferable or desirable in relation to the point of departure. This is the basis for the development of innovations, particularly from science. Alas, Sicard (2021) reminds us of the steady decline of the French industrial sector since the end of the Trente Glorieuses (30 years following the end of World War Two in France). This decline is illustrated in Fourastié (1952) and was followed by many governments because it seemed more profitable to them to move toward a post-industrial society consisting solely of service activities. It is only recently that the state has changed its "doctrine", leaving a more appropriate place for technological innovation.

In this context, the French see free trade, globalization and technological change as threats to jobs. Thus, according to Blanchard and Tirole (2021), manufacturing accounts for only 10.4% of employment and 13.4% of GDP in France, compared to 19.7% of GDP in Italy. Meanwhile, having lost 41% of manufacturing jobs in 30 years, the United States is one of the most active players in developing a new manufacturing-oriented vision (Japan and Germany have followed the opposite trend, going from 14.3% to 23.5% and 17.6% to 22.6%, respectively, according to Rynn (2010)). Industrial production is a key element of a country's wealth. According to Sicard (2022), there is a close correlation between a country's industrial production and the GDP/capita of its inhabitants. "Thus, with an industrial production of $6,432 per capita, France has a GDP/capita of 39,030 dollars; with a ratio of $12,279, Germany has a GDP/capita of $46,208; with a record ratio of $22,209, Switzerland is at a GDP/capita of 87,097 dollars". This situation, according to Andreoni and Gregory (2013), is based on, or calls for, the following arguments:

– The manufacturing industry is a crucial source of high-quality jobs, especially with the deployment of industry 4.0 principles and digital science and technology (André 2019).

– Around two-thirds of world trade is still in manually invoiced products (see balance of payments).

– The manufacturing industry is the main driver of economic growth because of its higher productivity and potential for innovation.

– "Human beings are far more capable than robots of dealing with unexpected situations" (Elon Musk, quoted by Alvarez (2018)).

– Technology has complex effects on the structure of production and on the distribution of labor: the replacement of labor by capital, and the elimination of low- and medium-skilled jobs with a high routine component (Blanchard and Tirole 2021).

For Andreoni and Gregory (2013), services only had a significant positive effect on jobs up until 1990, and with coefficients well below those of manufacturing. In the trends highlighted by the latter, one driver of progress in manufacturing is the production of machines using other machines. This may involve manufacturing our own machines (CECIMO 2011), but it is above all a matter of locally envisaging new processes, applicable in many fields (mechanical engineering, construction, computers, healthcare, automotive, aerospace, energy, space, metal fabrication, etc.), which make it possible to cut costs, improve quality and increase productivity, as well as reduce production set-up times.

Indeed, Bartik (2020) shows that financial incentives promote local job creation, as it takes a lot of money to sway a decision to locate or expand a business. So, even when there is a positive impact on employment, the budgetary costs per job preserved or created can be quite high. It can therefore be advantageous to rely on local creators without needing to attract companies whose ethics are sometimes questionable (major tax breaks granted to companies are not always appropriate).

The health crisis also revealed pre-existing problems: difficulties in terms of production and supply of consumer goods, the complexity (extreme fragmentation) of our production systems. Raffard (2021) wonders about the organization of these systems, with the multiplication of players involved at every stage. The supply, quality and accessibility of goods provide a privileged field of observation to reflect on the future evolution of our societies. This tendency toward prediction is as much "the projection of contemporary concerns, hopes and events onto the future" as it is a desire "to act on the present" (Belasco 2006).

But, according to Blanchard and Tirole (2021), on the one hand, state-financed and directed innovation programs lead to decisions on the type of innovations to be encouraged, and second, venture capital plays a relatively important role in financing short-term innovation (in the United States). It naturally favors fields offering investors the prospect of a rapid return on their investment. Lerner (2020) considers that there is a risk of excluding innovations that are profitable in the longer term or that could benefit society as a whole. For its part, China has an unrivalled ability to provide what experts call "a complete value chain", to the detriment of non-Chinese companies. Government guidance funds represent this interconnectivity in action, with global financial support of the order of US$700 billion (Blanchette and CSIS 2021; Luong et al. 2021). Meanwhile, in the West, an updated version of Luddism, an aversion to any job-limiting innovation, may counterbalance the new (Acemoglu and Restrepo 2019).

In several works (André 2017; Demoly and André 2022), we have shown that it is possible to continually change and personalize production and propose innovative solutions for production systems; in doing so, the innovations centered on additive manufacturing or 3D printing (or 4D if additive manufacturing is achieved with materials that can be activated by energy stimulation) act as "innovation bridges". In other words, they transfer their know-how and transform the way goods are produced and services are provided. We have just seen from Blanchard and Tirole (2021) that the direction and pace of technological change depend on many factors, many of which can be influenced by political decisions made by society and public authorities. So, are 4D printing modes, like 3D, among the chosen technologies? Or is it possible to show that, outside of major national or European programs, these emerging fields meet the right criteria for moving from ideas to their application in society?

Everyone knows, at least to a certain extent, that the choice of an already attractive technology depends on a large number of factors, such as production organization, expected gains, return on investment, corporate culture, the ability to hire trained staff and learning conditions for employees. Additive manufacturing – computer-aided manufacturing (CAM) – offers efficiency gains in material consumption and has significantly shortened the period between design and production, while other CAM systems allow increasing control over complex production processes (Mazzoleni 1997; Arnold 2001).

In addition, with the advent of IT, the operation of these productions can be interwoven with service activities and new technologies. The manufacturing industry is based on a "value chain" of activities ranging from basic research to after-sales service (see Figure P.1). This representation of the value chain highlights the potential interdependencies that can have a considerable influence on the possibilities of capitalizing on their innovations, capabilities and market opportunities. In the broadest sense, the aim of this book is to find out where these new manufacturing devices stand in terms of integration and/or strong competitive niches.

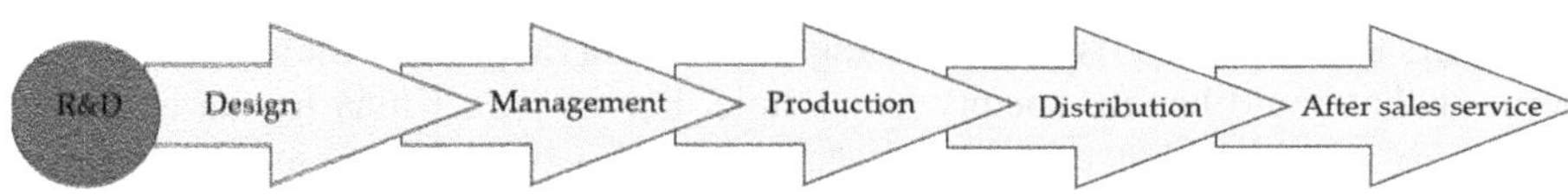

Figure P.1. *Interacting (and interdependent) elements in the transition from research to applications*

P.1. General thoughts on innovation

> [N]o latitude of expression would allow us to say [...] that the stream of water built the mill, though it were too ancient for us to know who the builder was. What the stream of water does in the affair, is neither more nor less than this; by the application of an unintelligent impulse to a mechanism previously arranged, [...] and arranged by intelligence, an effect is produced, viz. the corn is ground. But the effect results from the arrangement. The force of the stream cannot be said to be the cause or author of the effect [...]. There cannot be design without a designer; contrivance without a contriver; order without choice; arrangement, without any thing capable of arranging; subserviency and relation to a purpose, without that which could intend a purpose; means suitable to an end, and executing their office, in accomplishing that end, without the end ever having been contemplated, or the means accommodated to it (Paley 2006).

> In this context, the term "disruption" delineates a process in which new entrants with generally fewer resources challenge incumbent firms (Hopp et al. 2018, p. 446).

> A characteristic of modern societies is that they can only stabilize themselves dynamically, which means they need constant growth, acceleration and condensation of innovation to maintain [...] the status quo (Rosa 2021).

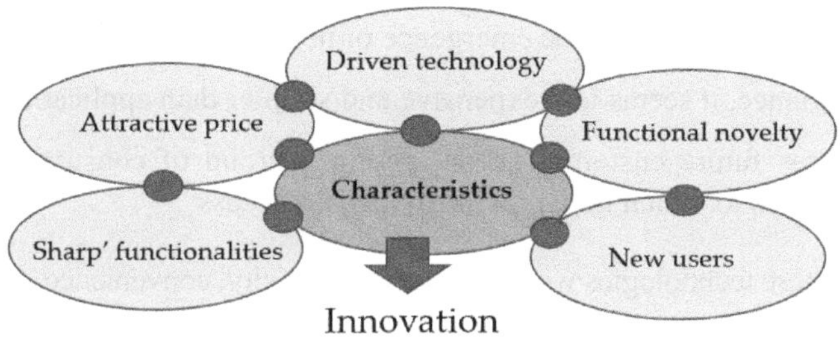

Figure P.2. *Characteristics of disruptive innovations*

When a radical innovation appears on the market, companies generally engage in a "low-attack" because it is promising, but not totally free from minor flaws. By starting out in niche markets, for example, performance can increase and the number of defects can decrease, before competing with incumbent operators or even finding new markets. As its quality improves, innovation begins to satisfy the requirements of traditional customers and replaces the technology. Figure P.2 (from Montgomery et al. (2018)) shows the characteristics of these disruptive innovations.

Do radical or disruptive innovations occur to create a more abundant world? What market conditions motivate innovators? Do they target communities suffering from poverty and resource scarcity? A theoretical model of disruptive innovation in a resource-constrained environment, integrating arguments from social capital theory, radical innovation, entrepreneurial action and social innovation can be built around the desire of citizens to access new and, if possible, low-cost technical products (Diamandis and Kotler 2012). As Mahto et al. (2020, pp. 6–7) argue:

> Companies that want to focus on innovation and specifically disruptive innovation should encourage non-conforming behavior in their employees and utilize various means to encourage creativity and individualism. In general, countries and companies should recognize

> the importance of social capital and devise ways to create conditions where social capital acts as a catalyst for disruptive innovation.

This is an original way of thinking that should be considered alongside other targets, particularly high-tech ones. For Zubizarreta et al. (2021), who take up this discourse towards low-tech, the concept of disruptive innovation has the following characteristics:

– characteristics that are undesirable among other user segments;

– technologies that are unsuitable for use in everyday applications;

– technologies that enable the emergence of new markets;

– at first glance, it seems less expensive and simpler than applications;

– anticipates future customer needs, giving a group of consumers access to a product or service to which they previously had no access.

Initially, these technologies would focus on simplicity, convenience, customizability and price, rather than results and high performance. But radical innovation is capable of creating entirely new and profitable markets or value propositions (Christensen 2013).

NOTE.– An example of innovation.

To take an example of an "illuminating" disruption, we consider lighting using the combustion of a thermal source (candle, kerosene lamp) and that provided by Edison's light bulb. The first requires oxygen, the other a vacuum; the first uses a local energy source, the other uses a distributed one (without having to get your hands dirty!); the first uses a combustible material, the other does not. Electric lighting represents a break away from the past.

P.1.1. *Toward proof of concept*

> Energy is a kind of fantastic projection; a projection which feeds all the industrial and technological dreams of modernity, and which also inflects the conception of Man in the direction of a dynamic of the will. However, we know from analyzing the phenomena of turbulence, chaos and catastrophe in the most recent physics that any flow, any linear process, when accelerated, takes on a strange curvature that is that of catastrophe (Baudrillard 1990).

> Too often, organizations, faced with potential disruption from competitors, commit resources to generating new ideas for exploration but struggle to convert these ideas into meaningful businesses (O'Reilly and Binns 2019).

> With Ricardo and Malthus, the notion of massive privation and great inequality became a basic premise. Smith's optimism gave way to a profound pessimism, resulting from the idea they had of man's relationship with nature and with each other (Galbraith 1961)[1].

The invention process covers all efforts to create new ideas and make them work. Irrespective of this important aspect of additive manufacturing, which I will come back to, society has the feeling that we are collectively accepting the irreversible, with major trends such as climate change and possible adaptations, no doubt on the margins. "Short-term emergencies are in direct contradiction with long-term constraints; inanity enters the realm of the possible, and this explains the depth of the malaise" (Barré et al. 2015). According to Theys (2003), whom they quote:

> In the case of environmental management, and in particular for global risks, the initial situation is one of poorly structured issues, numerous and poorly identified players, highly divergent interests, and unfavorable power relations. The conditions for "good governance" are not met *a priori*: they must therefore be partly constructed (Barré et al. 2015).

But, for these same authors, this look back to a time when these concerns were not the focus of sustained public attention is accompanied by new expectations. In this context, there is an important place for creativity, no doubt in a very different context than in the past.

Today, the interweaving between science and technology has "become so profound that we can now say that science has become as much a technician as a designer, since design itself very often depends on technical capabilities, and technical innovations are increasingly based on scientific knowledge" (Dubois and Brault 2021). Moreover, according to Collino (2022), "while the hallmark of all metaphysics and religions is to ensnare the individual in a network of beliefs enslaving him to absurd and criminal obligations; while they cloud the critical spirit, crystallize intelligence, foster ignorance, encourage and develop superstition, the materialist concept frees the human personality from these crushing servitudes, liberates it from fears and terrors, and shapes its judgment and reason". It is not a question of incremental following, but of creativity, questioning, using our critical minds, etc., to come up with something new and useful. This means that the diagram in Figure P.1 is normally under attack, because it assumes that science is the driving force, whereas it can be located in other places, such as in technology or social demand. However, in the action diagram, we need to equip ourselves with "rational

1 Author's translation from the French version of the text.

means of reliable ontological elaboration" (Bounge 2004), without renouncing objectivity.

To define a science policy, five conditions have been considered: a collective discourse that makes it explicit, political actors who take charge of it, public institutions that embody it and make it possible, achievements and results that represent it and legitimize it, and finally, a reflection on its construction and its stakes (Duclert 2006). But what about innovation? Should we follow Roqueplo (1983) when he writes: "Every technical object is a negotiation [...] it is simultaneously the contract that momentarily seals the balance of forces at play, and the weapon brandished for other conflicts and negotiations"? Do these new objects appear in a serene context or, on the contrary, despite the prevailing rhetoric, are they the result of a large body of new ideas that undergo a variety of filters in which the quality of people, financing, society, etc., all come into play in different ways? In this context, which could resemble the thinking expressed by Duclert, can we still confine ourselves to angelic speeches on innovation? Do we need to revisit the practices of research orientation and programming of research and its valorization based on the organization scientific debates and take them into account in public decision-making? The central question is that of the direction and trajectories of the creation of new objects or artifacts, transformed into innovations, which are the result of wills, initiatives and power relations. Before seeking to improve society's ability to grasp the invention-innovation aspect, it seemed important to examine the chain of interactions that exist between an idea and its transformation into a profitable invention, and thus create new knowledge.

This knowledge base can/should be used to better understand the difficult path that exists between the researcher, the laboratory (between individual initiatives and the organization's programed research orientations), and then between the many "support" mechanisms "to transform an idea into a technical proof-of-concept, then into a more "presentable" object, acceptable to society, finally proposed as an innovation...". No doubt reductively, we have introduced the "valley of death" concept "already used in this field", but in a broader sense. This apparently linear approach does not ignore the knowledge of elements in complex interactions and systems analysis in an interdisciplinary context.

To paraphrase Godard and Hubert (2003), we need to focus as much on the conditions required to ensure the success of research aimed at future, socially useful innovation, as on incremental research framed by French and European government programs "whose conditions and benchmarks are well defined, without being trapped by a discourse on scientific excellence [...] which would be at odds with the research needs raised by the issue". Thus, a partially conceptual and partially heuristic toolkit is revealed in an attempt to apprehend the relationships between these different stages, for which transaction zones exist, an in-between space in which modes of change are negotiated, coordinated and implemented.

However, common "intermediary languages" are needed at every stage, so that groups with different objectives and modes of understanding can exchange/communicate from the initial act of creation through to innovation. Galison (1997) speaks of "pidgins" or creoles when two human groups from different cultures use a simplified, pragmatically oriented third language to carry out an exchange. This situation of decompartimentation is an obligation that goes beyond the usual reductionism of social and technological determinism (Wyatt 2008). Indeed, as will be mentioned later, to translate this intercultural transition, we need to consider the logic of developing interdisciplinarity – that is all the discursive, organizational and architectural strategies put in place by research collectives with the aim of creating and maintaining a type of scientific work known as interdisciplinary (Li Vigni 2020).

This transfer context raises other questions, briefly mentioned below (see Hermesse and Vankeerberghen (2020)); first of all, it involves analyzing this reflection by Dewey (2014) for action:

> Theory, when separated from concrete action and doing, is empty and vain [...]. The problem of the relationship between theory and practice is not just a theoretical one. It is also the most practical problem we face in life. A problem that raises the question of how intelligence can inform action, and how action can enable a greater understanding of meaning: a clear perception of the values worthy of interest, and the means by which they can be secured through experienced objects. The construction of ideals in general and their sentimental glorification are easy; to leave it at that, however, is to evade our responsibilities for rigorous thought and action.

As a reminder, in line with Gregor and Hevner (2013), four typical configurations are associated with innovation: design, in which known solutions are applied to known problems; adaptation, in which the use of known solutions is extended to a new problem (analogy); and improvement, in which the use of known solutions is extended to a new problem (analogy); improvement linked to new solutions for known problems; and invention which involves creating new solutions to new problems (Pascal and Rouby 2017). In this last activity, which falls within the scope of this book, several elements can be questioned:

– Why and how can we be creative in a sometimes unreceptive academic world? Is there a conceptual basis for creativity? What are the specific methodological challenges involved in transforming an idea into a coherent, rational concept? Does the research team authorize the production of a proof of concept? What is the usefulness of this type of production and added value compared with other forms of research?

– Creativity, innovation, design-thinking are not the exclusive domain of design professionals. These fields can be explored by novices. These new practices require new interactions between different milieus. How can we encourage them so as not to limit invention to the "professionals" of the idea?

– How can we find, select and exploit personalities who are masters of cognitive tasks requiring creativity and intuition to solve problems requiring great leaps of logical imagination?

– How can we break down disciplinary silos?

– In the final analysis, do the envisaged disruptions reflect what is really happening in society as a whole (Bessen et al. 2020)?

– Can we think and create in an emergency? In particular, is it possible to plan for the immense challenge of our future in a context of extreme uncertainty (health and social crises, international tensions, etc.), or should we be content with the spontaneous emergence of ideas that can be transformed into innovations?

– How can the rise of the hustle economy be seen as a triumph of technological innovation and human creativity (Dewey 2020)? What are the effects on creativity and on the responsibility of research? What kind of ethics are there for the inventor?

– Can we maintain the principles of scientific objectivation, which is carried out under the following conditions: independence of approach, validation of results and their interpretation by peers, confirmation of results (Lecointre 2018)?

– Can we accept, on the pretext of invention, an activity carried out to the detriment of the scientific rigor and distancing that should accompany it (Smaldino and McElreath 2016)?

– How do shared inventions and competitive innovations follow trajectories that allow them to overcome their singularities in order to effectively realize this sharing and competition (Vigezzi 2019)?

– Can/should we set up hybrid spaces where researchers from a variety of scientific and non-scientific backgrounds "confront and cooperate in defining problems and lines of research, as well as the resources required for these investigations" (Dalle-Nazébi 2008)? Should these structures be temporary or permanent?

– What links should be forged or maintained between civil society and the researcher, the research team or the organization to which it belongs? How can disruptive research disruption influence the researcher's posture and trajectory: between enthusiasm and disgust?

– How can we take into account the timeframes of each stakeholder in these transfers?

– Can we then accept a single "crumpled" time adapted to the transition that raises the question of the radical management of separate timeframes?

– What loss (or gain) of autonomy is associated with the creative process?

– Research laboratory managers need to create a form of corporate culture. For the latter to be an advantage, is the trust that needs to be established and maintained with researchers, so that their experiences can be shared with the aspirations of the hierarchy possible? And how?

– Do we have sufficient capacity in terms of inventors to meet the needs of industry and thus of society? Does the shortage of talent and the severe skills gap, which Gold (2021) argues is threatening sustainable growth, have a role to play in the operation of academic research labs? Do we have disruptive incubators to transform university training programs from the inside into a preventive force for the innovations to be enabled and supported?

– How can we stop, or even reduce, the pressure exerted on researchers, who are condemned to a certain degree of followership, reflected in the success rates (i.e. the proportion of candidates who obtain funding) in many public funding agencies (Conseil des académies canadiennes 2021)? Does hyper-competition serve creativity?

– To put forward operation management in the form of a project (projected management mode) as a mode of intentional determination of behaviors in relation to the choice and decision is not easy. How does this operative anticipation of a desired future (Boutinet 2012) come about?

– Who is in charge of this operation? "Because of the multiple interactions and feedback within the environment in which it takes place, the action, once triggered, often escapes the actor's control, provoking effects that are unexpected and sometimes even contrary to those he expected" (Morin 2004).

– What are the reasons for justifying and rationalizing the action to be taken (ungrateful, inconsistent, etc.)?

– With a consistent creative process within a research group, is there evolution in the choice of its "horizons", in the determination of critical thresholds for revisited action, and in the consideration of expected effects in the "right timing" (concepts of adaptation, flexibility, adaptability, etc.)?

– Can the action scenarios concerning invention be reduced to the one envisioned by the researcher-designer (Carroll 2016)? Should they be extended to their organizational component?

– Are there any creativity training courses? Between training courses in which concepts are not the subject of questions, but enable rapid and effective learning, and more open paths involving creativity and innovation? What is the right choice? Are we mastering divergence/convergence (NAP 2014)?

– How can we value, evaluate, legitimize and "monetize" an experience where creativity results in a profitable application (including that resulting from the historical analysis of (possible) failures)?

– Should we follow the advice of Jollivet (2008a, 2008b): "There is no model; we're in the DIY business. With all that this word implies in terms of compromise, in relation to the objective pursued, imposed by local constraints of all kinds. But also with the need for creativity that this implies"?

– Generally speaking, should (can) we follow the paradigms of linear invention (Robbins 1932; Solow 1956), triangular invention (according to the associations between technical change and invention introduced by Freeman in 1984), the more innovation- and information-oriented paradigm proposed by Nelson and Winter (1982), the "multiple equilibria" paradigm of von Hippel (1986), the approach inspired by Science Studies developed by Picon (1995), or other forms (or even a mix of pre-existing and new elements proposed by Amendola and Gaffard (1988))? Do we need them?

– Finally, why is there no French Elon Musk (Silverzahn 2021)?

In this very broad context, questions concerning the theme of creativity, its translation into the most coherent proposal possible, and the realization of a proof of concept (POC) are posed in human, financial, organizational and other terms. The list of remarks and questions just presented already illustrate the richness of the general theme, which will be the subject of Chapters 1 and 2 of the present volume and of Chapter 1 of Volume 2 of the book, which envisages descriptions and reflections on the global innovation process. It also highlights the "valleys of death". It is a lengthy sequence, a veritable "chemin de croix" for the creative individual who wants to see their idea or discovery through to the end. This list will be expanded in the next section.

P.1.2. *Toward innovation*

> Social action refers to activities carried out by individuals, groups of individuals or collectives who form projects, i.e. who anticipate, in a more or less rational (i.e. thought-out and adapted) way, ends and the means to achieve them, and who, to carry out these projects, enter into relationships with one another. Social action produces more or less integrated, interlocking or coupled social systems. It brings into play

> phenomena of power present in all social domains (religious, political, intellectual, etc.) and therefore in the economic sphere, considerations of integrationand interrelations between agents considered as decision-makers and project agents. Decision-makers' projects are formed in a more or less intuitive or calculated way, taking into account the relationships and power that characterize these relationships (Perroux 1973).

> Perfect certainty is what Man wants. But it cannot be found in practical action and production. These occur in the horizon of an uncertain future, and involve perils, the risk of misadventures, frustration and failure (Dewey 2014).

> It seems all our current problems were generated by the solutions to problems of an earlier age. These new problems generated by the new invention are the source of the law of unintended consequences (Bonasso 2007).

We are moving into a different space here: inventions are promising ideas for products or processes, while innovations are commercialized products or processes (Chandy et al. 2006). Most of the scientific and technical activities organized in companies go beyond the "simple" generation of ideas (which may be exogenous), which do not systematically produce radical disruptions, but rather a broad base of progressive technological advances, sometimes leading cumulatively to major technical changes. These may be incremental or disruptive innovations.

"An innovation is an event (Mast 2006). It signifies a break in the routine order of things, a disruption that is then reintegrated into a progressively reconfigured understanding of routine. This disruption is semantic, but it is also 'real' insofar as the interpretive dimensions of culture draw the contours of the limits of material objects, because materiality is impregnated and constituted by meaning" (Mast et al. 2013). Schön (1967) pointed out that innovation can occur when, like plate tectonics, cultural concepts and systems of understanding of the past collide with those of the present. In such cases, our understandings are simultaneously based on historical concepts and the disruption induced by the new situation. After a translation in the form of an idea to be promoted, the envisaged research leading to invention translates into the creation of POC and written proof in the form of patent applications.

In addition, this operation involves technological work, often carried out in partnership and interdisciplinary work. While, on the one hand, there is the cross-fertilization of knowledge and its pooling, and thus cultural enrichment through exchange, the POC remains an artifact demonstrating technical feasibility based on science. This initial research leads to reflexivity on the part of the institution that hosts the designer, on what could make it evolve to encourage invention and principles of co-creation with a social perspective on POC (interdisciplinarity and social sciences). *Design* contributes to innovation and business development (Bertola and Teixeira 2003; Battistella et al. 2012) because of its goals at the place of orientations, strategic vision and iterative methodologies, combined with its creative thinking in relation to technology (Nussbaum 2011; Glaubert and Bergeron 2020).

It is obvious that we design for users, but why not work with them to define everyday objects on a technical basis, but adapted to their use (ergonomics, cost, etc.). Co-construction is obviously not a label to be placed on the artifact, just a slogan. In this interactive context, co-construction is attentive to the needs of those for whom it is intended and who will be its users. By adding style to functionality, the designer testifies to their consideration for them. If this direction of action is in continuity with what was stated in section P.1, where the idea sprang from the head of a scientist (no doubt because the author is a member of that tribe!), it can just as easily come from the social body. In this case, we are talking about demand-side management.

Under these conditions, it is a matter of linking demand with those who can satisfy it, that is, in the first instance, technicians and, if necessary, scientists. The problems of filtering (the "valleys of death") already mentioned are found again:

– How can we get an idea off the ground and take it into consideration?

– Do we have a receptive industrial base?

– Does the co-creation group represent the social component concerned by the artifact? Will we go beyond invention? Or even just the idea or its beginnings?

– Is the role of the departments responsible for marketing new products appropriate?

The word "creativity" is often used to regain control over economic performance and emerge from various financial crises. The prevailing discourse advocates this potential with little explanation of how and why an economic approach to creativity is useful and necessary. If it is useful to understand how creative people work and act in an uncertain and changing universe (Burger-Helmchen 2013), do we have any good recipes to encourage their creative productivity? Second, it would be useful to understand how organizations integrate and exploit creative ideas.

To achieve these goals, learning and research processes are essential. But new knowledge and artifacts lead to changes in both the way we do things and the way we are, and as a result, we are faced with change. Yet, "we develop familiar patterns of behavior that define a successful and functional daily life in the world, and we find it difficult to change our habits" (Bonasso 2007). For this author, discovery requires people who are curious about what is new and novel, and dissatisfied with the status quo:

> These are people who are curious and who ask old questions in new ways and who ask new questions. Inventions are created by curious people dissatisfied with current methods and means of working. These are people who seek something more than an incremental refinement in the usual way of working. And, innovation is spawned by dissatisfied people looking for new opportunities behind doors believed by most people to be closed (Bonasso 2007).

The possible users of the results of ideas transformed into objects delimit boundaries of interpretation that can dictate by whom, how and for what they can be used. Sometimes, it takes several years before an invention is accepted by the public. This is another "valley of death" to consider (see, for example, Sharkey (2010)).

But, after proposing an idea invention, for many, a lack of understanding of the world of research and engineering is obvious; it can give way to doubts and suspicions of conflicts of interest, media and, above all, economic influences. Invention may be almost free, but well-conducted innovation can result in benefits of a different magnitude. The idea of "piracy" is not far off! In this environment, which can be distrustful and uncertain, guaranteeing trust in the partnership is essential. Now, more than ever, making an object "civilized" with uses adapted by and for society is a necessity in a context where, in a complementary way, we need to design products and services capable of responding to the major present and future societal challenges. So, let us not make too many mistakes! For example, the "valley of death" problem may be more serious for companies in emerging markets than in mature markets, depending on the products or processes involved (Zhou and Wang 2020).

According to Burger-Helmchen (2013), innovation increasingly involves rapid, short-term returns on investment. At the same time, for reasons of economic fragmentation, we are witnessing an erosion of technical capabilities and, in global competition, a possible gradual destruction of the competitiveness of products and processes (Aghion et al. 2020). In an attempt to maintain a robust innovation structure, strong and positive relationships between R&D and marketing organizations could significantly improve the relevance of innovations when new products are introduced (provided that a suitable communication language exists).

As we can see, innovation is an all-purpose word that simply translates the translation of an invention to the market, with many influencing factors that will need to be analyzed in greater depth.

P.1.3. *Between idea, invention and innovation*

> They [the guilds] pass on useful techniques, but block innovations that pose a threat. Guilds largely existed, not because they corrected market failures or served the common good, but because they benefited two powerful groups: guild members and political elites (Ogilvie 2019)[2].

> Forget short-term symptoms like Donald Trump and Brexit, it's innovation that will shape the 21st century. Yet innovation remains a mysterious process, poorly understood by policymakers and businesspeople alike (Ridley 2020)[3].

> Transforming basic research discoveries into marketable products or technologies is a major challenge that requires individuals to have a specific set of skills. Among these skills, the ability to "translate", particularly between basic and applied research, and to build bridges between these fields, is likely to be decisive (Assmus and Haeussler 2017)[4].

"Discontinuous innovation involves establishing a new product and setting up new models of behavior" (Robertson 1967). Roberts and Frohman (1978) and Boly et al. (2016) have shown the existence of an imbrication of numerous elements that can enable the emergence of new processes or artifacts. Until recently, approaches to innovation were able to develop in the logic of supply, with a break between a technical proposal and society. Open innovation, a more recent concept, is the antithesis of the previous model, which is a model of vertical integration in which the activities related to innovation are internal to the design world. This paradigm, introduced by Chesbrough (2003) and Chesbrough et al. (2006), can develop through interactive processes, via a logic of networks, assuming that ideas come from both outside and within culturally separated worlds (Badillo 2013).

Many successful technical innovations seem to have been initiated by activities responding to a "market pull" (see Bianchini et al. (2019)), but, unexpectedly, other innovations result from the recognition of a market need or opportunity. Sometimes, according to Roberts (2007), innovation results from a highly motivated individual disrupting the market's consciousness until it agrees to launch and support an original technical program (especially when associated with the company's competence). The growth of research consortiums and strategic alliances between

2 Author's translation from the French version of the text.

3 Author's translation from the French version of the text.

4 Author's translation from the French version of the text.

companies and *startups* in emerging technologies is also an indicator of the search for innovations, disruptive if possible. In this expanding landscape, Fab-Labs also have their place (Damoah and Botchie 2020). Finally, Gambardella et al. (2012) have shown that users create and drive innovations for their own use, followed by the adoption of these innovations by equipment manufacturers, etc. (see Figure P.3).

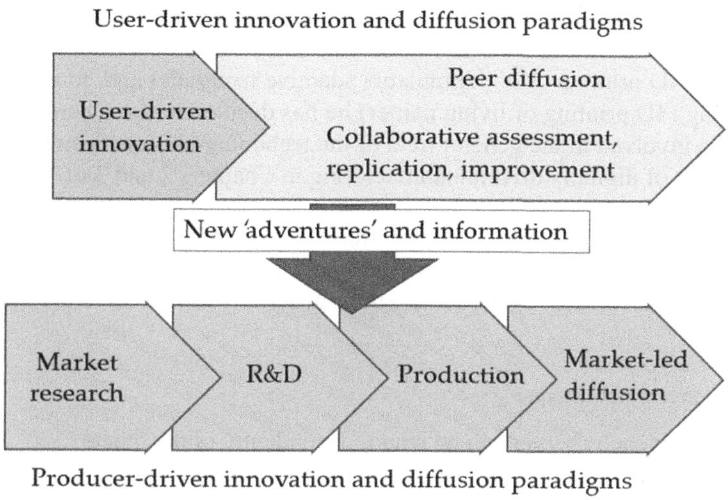

Figure P.3. *Innovation between user and producer*

P.2. The example of 3D, 4D and bio-printing manufacturing

> Ontologies represent the essential technology that enables and facilitates interoperability at the semantic level, providing a formal conceptualization of the data that can be shared, reused and aligned (D'Aquin and Noy 2012).

> Short of a time machine, science is the most powerful tool that humanity has invented to bring the future into the present (Nowotny 2015).

> Innovation is also believed to be predictive of successful scientific careers: Innovators are science's trailblazers and discoverers, so producing innovative science may lead to successful academic careers (Hofstra et al. 2020).

Nowotny (2015) warns us, those who succeed remind us that innovation is a highly uncertain process that is not only the work of intelligent, determined individuals, but also of numerous interdependent processes, integrated into an ecosystem that feeds on contingencies. Faced with such a spectrum, it is impossible to draw credible directions or lines of force without having put a certain number of key criteria to the test, ranging from the idea, imagination, invention, to application. The author has been familiar with 3D printing (referred to as additive manufacturing (André et al. 1984)) since the first patent in 1984. Because of his involvement in research on 4D printing (3D + stimulable adaptive materials) and, to a lesser extent, bio-printing (4D printing of living matter) he has decided to test a certain number of the criteria involved in the general field of the technological enhancement of thought on the theme of digitally driven manufacturing in Chapters 2 and 3 of Volume 2.

Some robust conclusions emerge from this comparison between theoretical reflections and their testing on themes is considered to be disruptive. They will be discussed and presented; in particular, their generality will receive the author's full attention.

NOTE.– General comments

1) Much of each chapter can be read independently of the others.

2) When it comes to men and women, Man is written with a capital M.

November 2023

P.3. References

Acemoglu, D. and Restrepo, P. (2019). The wrong kind of AI? Artificial intelligence and the future of labor demand. *Cambridge Journal of Regions, Economy and Society*, 13, 25–35.

Aghion, P., Antonin, C., Bunel, S. (2020). *Le pouvoir de la destruction créatrice.* Odile Jacob, Paris.

Alvarez, S. (2018). Inside Tesla's "tent"-based model 3 line that set a path to profitability. *Teslarati*, 10 December [Online]. Available at: https://www.teslarati.com/inside-tesla-tent-based-model-3-production-assembly-line-profitability/.

Amendola, M. and Gaffard, J.L. (1988). *La dynamique économique de l'innovation.* Economica, Paris.

André, J.C. (2017a). *From Additive Manufacturing to 3D/4D Printing 1: From Concepts to Achievements.* ISTE Ltd, London, and John Wiley & Sons, New York.

André, J.C. (2017b). *From Additive Manufacturing to 3D/4D Printing 2: Current Techniques, Improvements and their Limitations.* ISTE Ltd, London, and John Wiley & Sons, New York.

André, J.C. (2017c). *From Additive Manufacturing to 3D/4D Printing 3: Breakthrough Innovations: Programmable Material, 4D Printing and Bio-printing*. ISTE Ltd, London, and John Wiley & Sons, New York.

André, J.C. (2019). *Industry 4.0: Paradoxes and Conflicts*. ISTE Ltd, London, and John Wiley & Sons, New York.

André, J.C., Le Méhauté, A., De Witte, O. (1984). Dispositif pour réaliser un modèle de pièce industrielle. French patent no. 84 11 241.

Andreoni, A. and Gregory, M. (2013). Why and how does manufacturing still matter: Old rationales, new realities. *Revue d'économie industrielle*, 144, 21–57.

Arnold, H. (2001). The recent history of the machine tool industry and the effects of technological change. LMU Working paper 14 [Online]. Available at: https://www.researchgate.net/publication/228947863_The_recent_history_of_the_machine_tool_industry_and_the_effects_of_technological_change.

Assmus, A. and Haeussler, C. (2017). At the interface between invention and innovation: Increasing success rates in clinical trials. *Academy of Management Annual Meeting Proceedings*, Boston.

Badillo, P.Y. (2013). Les théories de l'innovation revisitées : une lecture communicationnelle et interdisciplinaire de l'innovation ? Du modèle "Émetteur" au modèle communicationnel. *Les Enjeux de l'information et de la communication*, 14, 19–34.

Barré, R., Lavoux, T., Piveteau, V. (eds) (2015). Trop tôt, trop tard ? In *Un demi-siècle d'environnement entre science, politique et prospective – En l'honneur de Jacques Theys*. Quæ, Paris.

Bartik, T.J. (2020). Bringing jobs to people: Improving local economic development policies. Policy paper no. 2020-023, W.E. Upjohn Institute for Employment Research. doi: 10.17848/pol2020-023.

Battistella, C., Gianluca, B., De Toni, A.F. (2012). From design driven innovation to meaning strategy. *Management Decision*, 50, 718–743.

Baudrillard, J. (1990). *La transparence du mal*. Galilée, Paris.

Belasco, W. (2006). *Meals to Come – A History of the Future of Food*. University of California Press, Berkeley.

Bertola, P. and Teixera, J.C. (2003). Design as a knowledge agent. *Design Studies*, 24, 181–194.

Bessen, J.E., Denk, E., Kim, J., Righi, C. (2020). Declining industrial disruption. Research paper 20-28, Boston University [Online]. Available at: http://dx.doi.org/10.2139/ssrn.3682745.

Bianchini, S., Llerena, P., Patsali, S. (2019). Demand-pull innovation in science: Empirical evidence from a research university's suppliers. *Research Policy*, 48, 100005.

Blanchard, O. and Tirole, J. (eds) (2021). Les grands défis économiques [Online]. Available at: https://www.strategie.gouv.fr/sites/strategie.gouv.fr/files/atoms/files/fs-2021-rapport-les_grands_defis_economiques-juin_0.pdf.

Blanchette, S. and CSIS (2021). Confronting the challenge of Chinese state capitalism [Online]. Available at: https://www.csis.org/analysis/confronting-challenge-chinese-state-capitalism.

Boly, V., Camargo, M., Morel, L. (eds) (2016). Définitions de l'innovation technologique. In *Ingénierie de l'innovation*. Lavoisier, Paris.

Bonasso, S.G. (2007). Inquiry, discovery, invention, and innovation – The personal experience of technology generation and transfer in engineering and scientific research. *Leadership and Management in Engineering*, 7, 141–150.

Boutinet, J.P. (2012). *Anthropologie du projet*. PUF, Paris.

Bunge, M. (2004). *Matérialisme et humanisme : pour surmonter la crise de la pensée*. Liber, Montreal.

Burger-Helmchen, T. (2013). *The Economics of Creativity: Ideas, Firms and Markets*. Routledge, London.

Carroll, J.M. (2016). Theorizing the unprecedented. In *Theory Development in the Information Sciences*, Sonnewald, D.H. (ed.). University of Texas Press, Austin.

CECIMO (2011). Study on competitiveness of the European machine tool industry [Online]. Available at: https://www.cecimo.eu/publications/study-on-the-competitiveness-of-the-european-machine-tool-industry/.

Chandy, R., Hopstaken, B., Narasimhan, O., Prabhu, J. (2006). From invention to innovation: Conversion ability in product development. *Journal of Marketing Research*, 43, 494–508.

Chesbrough, H. (2003). *Open Innovation: The New Imperative for Creating and Profiting from Technology*. Harvard Business School Press, Boston.

Chesbrough, H., Vanhaverbeke, W., West, J. (eds) (2006). *Open Innovation: Researching a New Paradigm*. Oxford University Press, Oxford.

Christensen, C.M. (2013). *The Innovator's Dilemma: When New Technologies Cause Great Firms to Fail*. Harvard Business School Press, Boston.

Collino, R. (2022). Entrée – Matérialisme. In *L'Encyclopédie anarchiste*, Faure, S. (ed.). Paris [Online]. Available at: http://www.encyclopedie-anarchiste.org/.

Conseil des académies canadiennes (2021). Dynamiser la découverte [Online]. Available at: https://www.rapports-cac.ca/wp-content/uploads/2021/05/Dynamiser-la-decouverte-Rapport_FR_DIGITAL_FINAL.pdf.

D'Aquin, M. and Noy, N.F. (2012). Where to publish and find ontologies? A survey of ontology libraries. *Journal of Web Semantics*, 11, 96–111.

Dalle-Nazébi, S. (2008). Objet et acteurs de recherche – La montée en expertise de locuteurs de langues des signes. *Revue d'anthropologie des connaissances*, 2, 63–96.

Damoah, I.S. and Botchie, D. (2020). Do-It-Yourself (DIY) laboratories and science, technology, and innovation (STI): Trends, implications and future research. *Technology Analysis & Strategic Management*, 33, 1267–1280.

Demoly, F. and André, J.C. (2022a). *4D Printing 1: Between Disruptive Research and Industrial Applications*. ISTE Ltd, London, and John Wiley & Sons, New York.

Demoly, F. and André, J.C. (2022b). *4D Printing 2: Between Science and Technology*. ISTE Ltd, London, and John Wiley & Sons, New York.

Dewey, J. (2014). *La quête de certitude : une étude de la relation entre connaissance et action*. Gallimard, Paris.

Dewey, C. (2020). The gig economy is failing. Say hello to the hustle economy [Online]. Available at: https://onezero.medium.com/the-gig-economy-is-failing-say-hello-to-the-hustle-economy-13ae3aa91954.

Diamandis, P. and Kotler, S. (2012). *Abundance: The Future Is Better Than You Think*. Free Press, New York.

Dubois, M.J.F. and Brault, N. (2021). *Manuel d'épistémologie pour l'ingénieur*. Éditions Matériologiques, Paris.

Duclert, V. (2006). La France et la politique de recherche au xxe siècle. L'enjeu historique de l'institutionnalisation. In *Le gouvernement de la recherche. Histoire d'un engagement politique, de Pierre Mendès France au Général de Gaulle (1953-1969)*, Chatriot, A. and Duclert, V. (eds). La Découverte, Paris.

Fourastié, J. (1952). *Le grand espoir du XXe siècle*. PUF, Paris.

Freeman, C. (1984). L'économie de la recherche et du développement. In *Les enjeux du progrès : science, technologie et société*, Cambrosio, A. and Duchesne, R. (eds). Presses de l'Université du Québec/Télé-Université, Quebec.

Galbraith, J.K. (1961). *L'ère de l'opulence*. Calmann-Lévy, Paris.

Galison, P.L. (1997). *Image and Logic – A Material Culture of Microphysics*. University of Chicago Press, Chicago.

Gambardella, A., Raasch, C., von Hippel, E.A. (2012). The user innovation paradigm: Impacts on markets and welfare. *Management Science*, 63, 1450–1468.

Glaubert, D. and Bergeron, L. (2020). Le design comme accélérateur de projets technologiques en phase d'incubation précoce. *Sciences du Design*, 12, 78–91.

Godard, O. and Hubert, B. (2003). Le développement durable et la recherche scientifique à l'INRA [Online]. Available at: http://www.inra.fr/developpement-durable/RapportDevDurable.pdf.

Gold, S. (2021). Let's prepare graduates for jobs that don't exist (yet) [Online]. Available at: https://www.industryweek.com/talent/education-training/article/21164020/lets-prepare-graduates-for-jobs-that-dont-exist-yet.

Gregor, S. and Hevner, A.R. (2013). Positioning and presenting design science research for maximum impact. *MIS Quarterly*, 37, 337–356.

Hermesse, J. and Vankeerberghen, A. (2020). La recherche transdisciplinaire au sein des institutions d'enseignement supérieur et de recherche. *Natures Sciences Sociétés*, 28, 270–277.

Hofstra, B., Kulkarni, V.V., Munoz-Najar, G.S., He, B., Jurafsky, D., McFarland, D.A. (2020). The diversity-innovation paradox in science. *Proceedings of the National Academy of Sciences – PNAS*, 117, 9284–9291.

Hopp, C., Antons, D., Kaminski, J., Salge, T.O. (2018). Disruptive innovation: Conceptual foundations, empirical evidence, and research opportunities in the digital age. *Journal of Product Innovation Management*, 35, 446–457.

Jollivet, M. (2008a). Présentation. *Natures Sciences Sociétés*, 16, 4–6. doi: 10.1051/nss:2008047.

Jollivet, M. (2008b). Compte-rendu des Journées. *Natures Sciences Sociétés*, 16, 7–14. doi: 10.1051/nss:2008047.

Lecointre, G. (2018). *Les sciences face aux créationnismes – Réexpliciter le contrat méthodologique des chercheurs*. Quæ, Versailles.

Lerner, J. and Ramana, N. (2020). Venture capital's role in financing innovation: What we know and how much we still need to learn. *Journal of Economic Perspectives*, 34, 237–261.

Li Vigni, F. (2020). Architecture et interdisciplinarité. Le cas d'étude des instituts de la complexité. *Natures Sciences Sociétés*, 28, 306–313.

Luong, N., Arnold, Z., Murphy, B. (2021). Understanding Chinese government guidance funds. Report, Center for Security and Emerging Technology. doi: 10.51593/20200098.

Mahto, R.V., Belousova, O., Ahluwalia, S. (2020). Abundance – A new window on how disruptive innovation occurs. *Technological Forecasting and Social Change*, 155, 119064.

Mast, J.L. (2006). The cultural pragmatics of eventness. In *Social Performance: Symbolic Action, Cultural Pragmatics, and Ritual*, Alexander, J., Giesen, B., Mast, J.L. (eds). Cambridge University Press, Cambridge.

Mast, J.L., Ahrweiler, P., Viale, R. (2013). Cultural theory and its spaces for invention and innovation. *Mind & Society*, 12, 23–33.

Mazzoleni, R. (1997). Learning and path-dependence in the diffusion of innovations: Comparative evidence on numerically controlled machine tools. *Research Policy*, 26, 405–428.

Mitcham, C. (1998). The importance of philosophy to engineering. *Teorema*, 17, 27–47.

Montgomery, N., Squires, G., Syed, I. (2018). Disruptive potential of real estate crowdfunding in the real estate project finance industry – A literature review. *Property Management*, 36, 597–619.

Morin, E. (2004). *La méthode – Éthique*. Le Seuil, Paris.

NAP (2014). Convergence: Facilitating transdisciplinary integration of life sciences, physical sciences, engineering, and beyond [Online]. Available at: http://nap.edu/18722.

Nelson, R.R. and Winter, S.G. (1982). *An Evolutionary Theory of Economic Change*. Belknap Press & Harvard University Press, London.

Nussbaum, B. (2011). Designers are the new drivers of American entrepreneurialism [Online]. Available at: https://www.fastcompany.com/90184041/designers-are-the-new-drivers-of-american-entrepreneurialism.

O'Reilly, C. and Binns, A.J.M. (2019). The three stages of disruptive innovation: Idea generation, incubation and scaling. *California Management Review*, 61, 49–71.

Ogilvie, S. (2019). *The European Guilds: An Economic Analysis*. Princeton University Press, Princeton.

Paley, W. (2006). *Natural Theology: or, Evidences of the Existence and Attributes of the Deity*, 12th edition. Darwin Online, London [Online]. Available at: http://darwin-online.org.uk/converted/Ancillary/1809_Paley_A142.html.

Pascal, A. and Rouby, E. (2017). Les scénarios d'usage comme support aux méthodologies de Recherche en Design Science dans le cas d'invention. *Systèmes d'information & management*, 22, 47–78.

Perroux, F. (1973). *Pouvoir et économie*. Bordas, Paris.

Picon, A. (1950). Construction sociale et histoire des techniques. *Annales, Histoire, Sciences Sociales*, 3, 531–535.

Raffard, P. (2021). *Géopolitique de l'alimentation et de la gastronomie. De la fourche à la FoodTech*. Le Cavalier Bleu, Paris.

Richet, C. (1895). La science a-t-elle fait banqueroute ? *Revue Scientifique*, 3, 33–39.

Ridley, M. (2020). *How Innovation Works: And Why It Flourishes in Freedom*. Harper & Collins Publishers, New York.

Robbins, L. (1932). *An Essay on the Nature and Significance of Economic Science*. Macmillan, London.

Roberts, E.W. (2007). Managing invention and innovation: Technological innovation can alter the competitive status of firms and nations but its purposeful management is complex, involving the effective integration of people, organizational processes and plans. *Research Technology Management*, 50, 35–54.

Roberts, E.W. and Frohman, L. (1978). Strategies for improving research utilization. *Technology Review*, 80, 32–39.

Robertson, T.S. (1967). The process of innovation and the diffusion of innovation. *Journal of Marketing*, 31, 14–19.

Roqueplo, P. (1983). *Penser la technique : pour une démocratie concrète*. Le Seuil, Paris.

Rosa, H. (2021). *Remède à l'accélération*. Champs Essais, Paris.

Rynn, J. (2010). *Manufacturing Green Prosperity: The Power to Rebuild the American Middle Class*. Praeger Press, New York.

Schön, D.A. (1967). *Invention and the Evolution of Ideas*. Tavistock, London.

Sharkey, J. (2010). Reinventing the suitcase by adding the wheel [Online]. Available at: http://www.nytimes.com/2010/10/05/business/05road.html.

Sicard, C. (2021). L'industrie : un sujet pour juger de la pertinence des candidats [Online]. Available at: https://www.contrepoints.org/2021/12/24/417870-l-industrie-un-sujet-pour-juger-de-la-pertinence-des-candidats?mc_cid=3ae7f77129&mc_eid=1b4a9d2e39.

Sicard, C. (2022). Désindustrialisation : France Stratégie se plante dans ses prédictions [Online]. Available at: https://www.contrepoints.org/2022/03/16/423367-desindustrialisation-france-strategie-se-plante-dans-ses-predictions.

Silberzahn, P. (2021). Ce qu'un vieux dessin animé nous dit du manque d'innovation en France [Online]. Available at: https://www.contrepoints.org/2021/11/12/413008-ce-quun-vieux-dessin-anime-nous-dit-du-manque-d-innovation-en-france.

Smaldino, P.E. and McElreath, R. (2016). The natural selection of bad science. *Royal Society Open Science*, 3, 160384.

Solow, R.M. (1956). A contribution to the theory of economic growth. *Quarterly Journal of Economics*, 70, 65–94.

Theys, J. (2003). La gouvernance, entre innovation et impuissance : le cas de l'environnement. *Développement durable et territoires*. doi: 10.4000/developpementdurable.1523.

Valéry, P. (1957). *Œuvres*. Gallimard, Paris.

Vigezzi, M. (2019). Les inventions partagées et les innovations concurrentielles : des approches nouvelles. *Innovations*, 59, 243–260.

Von Hippel, E. (1986). Lead users: A source of novel product concepts. *Management Science*, 791–805.

Wyatt, S. (2008). Technological determinism is dead; long live technological determinism. In *The Handbook of Science and Technology Studies*, Hackett, E.J., Amsterdamska, O., Lynch, M., Wajcman, J. (eds). MIT Press, Boston.

Zhou, C. and Wang, R. (2020). From invention to innovation: The role of knowledge-intensive business services in technology commercialization. *Technology Analysis & Strategic Management*, 32, 1436–1448.

Zubizarreta, M., Ganzarain, J., Cuadrado, J., Lizarralde, R. (2021). Evaluating disruptive innovation project management capabilities. *Sustainability*, 13, 1. doi: 10.3390/su13010001.

Introduction

The essential point is that the incentive to make an invention, like the incentive to produce any other good, is affected by the excess of expected returns over expected costs.

(Schmookler 1962)

This [the invention] represents a new situation, the creation of a technical or organizational novelty, concerning goods, services or devices, whereas innovation represents the whole social and economic process leading to the invention being finally used, or not.

(Alter 2002)

The event undoes the temporality required to invent another time, from which another world, another way of looking at things, is inaugurated.

(Dufourmentelle 2011)

"Eureka, I've got it! Eureka"[1] is an interjection that would have been used to celebrate a discovery or invention by Archimedes. It is a transliteration of an exclamation generally attributed to this Greek mathematician and inventor.

> These events, snapshots of discovery in the imagination, show that discovery is the basis of a moment of existential tension and questioning, generating emotional behavior. This occurs at an implicit level of consciousness, an infra-conscious level where emotion is felt through immediate bodily involvement, without being grasped by consciousness: experience is lived without any conscious decision thought through gestures, postures, sensations, cries, sounds, homeostatic adjustments, i.e. biochemical regulations that maintain the equilibrium of the person's inner environment (Benedetti 2021).

1 See: https://en.wikipedia.org/wiki/Eureka_(word).

But, as Valéry (1957) writes:

> Eureka taught me in a few moments Newton's law, Laplace's name, the hypothesis he proposed, the very existence of research and speculation that we never told teenagers about, lest, I imagine, they become interested.

The world has changed in this respect, and many (FMI 2021) consider that inventions today draw on pure scientific knowledge: "While applied research is important for bringing innovations to market, fundamental research broadens the knowledge base needed to achieve scientific breakthroughs". For Ridley (2020), the myth of the lone inventor needs to be challenged: there is no "Eureka!" technology. For this author, innovation (exploited/used by society) is part of an evolutionary process in which a large number of people are involved: it uses existing knowledge to combine it in original ways. What is more, inventions are often made almost simultaneously in different places, by different people (who don't know each other), even if, in the end, the inventor is the first to walk through the door of the patent office (or is American!):

> In the middle of the 20th century, creativity was established as an attribute of citizenship in industrialized countries. This was neither an empirical discovery resulting from observation, nor a theoretical construct designed to increase the rationality of our understanding of the world. The aim was clearly to make every individual, as it were, a potential source of originality and change. By thus wresting the citizen, at least in appearance, from the seriality of the masses, we could expect three types of gain: a gain in distributed identity, a gain in productivity and a gain in social recognition (Rouquette 2007).

So, if you are capable of inventing and creating something previously unthought of, is it not that a way of being an accomplished citizen and, in this case, having the power to positively signal yourself to the attention of others, and less conclusively, to that of your dear colleagues (Brassens is not far off, with the conformity of thought)? From a more personal point of view, it is not entirely untrue that the patent has constituted a reference: a solidly established reference. In 1984, the author of this book, along with two colleagues from CILAS (Compagnie industrielle des lasers) – Alain Le Méhauté and Olivier de Witte – filed the first additive manufacturing patent called laser stereo-photolithography. When we started working on a manufacturing process based on light-sensitive resin and light, it was above all an intellectual game exploiting, in an undoubtedly original way, pre-existing knowledge (which was later developed). More precisely, it was a challenge issued to me by Alain of CILAS, then a division of Alcatel located in Marcoussis (Essonne). Dortier (2015) wrote:

> Are innovations revolutionary? Are they based on originality, the surprising, the unexpected? On the contrary, most are very ordinary. They are based on simple, predictable laws of evolution that nature itself knows how to put to good use. Innovating is [just] thinking outside the box.

As I am a researcher specializing in light-matter interactions, Alain wanted to submit a problem he was working on to me. So, from the outset, the idea was not designing industrial equipment. However, when the first creations came out of our first stereo-lithography machine, which was very much a cottage industry, we immediately saw the potential of the process. There is something magical about seeing an object being built, layer by layer, before your very eyes. I think this is what attracted so many people to the process, and guided them toward other "sister" technologies to what is now integrated under the name of "additive manufacturing". 3D printing was developed on the principle of additivity in 1984 by playing on a simple matter-energy coupling, with stereo-lithography as its first technology, representing a one-photon photopolymerization process. Because of this principle, other processes have been able to develop, in particular, using solid, pasty, powdery, organic, metallic and/or mineral materials.

Before I joined the research project, my colleagues at CILAS were exploring a selective polymerization process using two-photon absorption. They had taken over the study of patents – in my view, purely theoretical, with no *proof of concept* – from the American Institute of Chemistry. The original idea was to make the material react with two laser beams, exploiting a nonlinear absorption process. With this technology (Adamson 1968; Adelman and Lewis 1971; McGinniss and Schwerzel 1984), printing does not take place layer by layer, but within the volume of the liquid resin bath. At the time, there were a number of technical difficulties with this process, in particular the lasers required were particularly expensive and cumbersome, the resins insufficiently transparent and so on. So, lacking the appropriate lasers, I decided to keep things simple and concentrate on shallow surface polymerization (see Figure I.1).

So, I borrowed some mono-reactive acrylic resins from the macromolecular chemistry-physics lab next door to mine, and it was possible to complete a proof of concept in 3 days. An important element in this little speech is the use of "I" and not "we". Indeed, stimulated by my two friends, I was able to concentrate all of my "interdisciplinary" activity on the 3D subject for a short period of time: there was no need to find helpers from fields other than my own, nor to set up an interdisciplinary project group to achieve a proof of concept, nor to look for a few subsidies to get ahead! The old dogma of invention *Keep It Simple* could not have had a better setting!

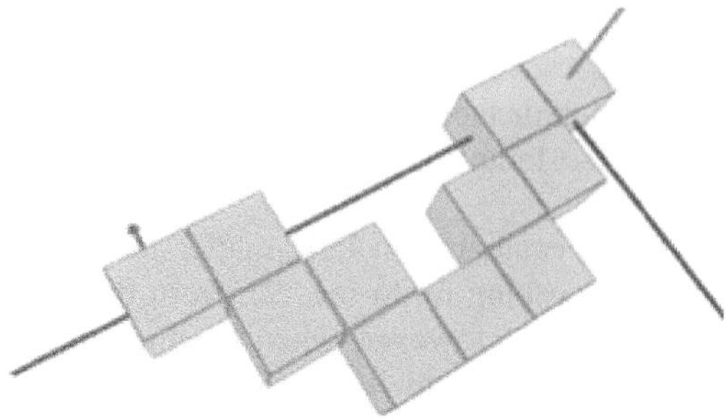

Figure I.1. *Construction by voxels (or elementary bricks). For a color version of this figure, see www.iste.co.uk/andre/knowledge1.zip*

On this basis, the object is built layer by layer (hence the term additive manufacturing). What I did not know in 1984 was that the presence of oxygen in the air played a role in inhibiting polymerization: each voxel in which oxygen was consumed became solid, but a small thickness (of the order of μm) remained fluid because the oxygen in the air had been able to diffuse onto the surface of the object. Drawing a parallel with masonry, bricks and their "paste" or cement were created simultaneously. *Serendipity* exists, even in a CNRS team! Indeed, in the second layer (see Figure I.2), air is no longer able to diffuse, leading to the polymerization of the resin, now unprotected by oxygen, and the bond between each layer. The notion of "printability" was thus born.

The very first versions therefore worked without a laser, by focusing the light of the sun! It may seem strange, but even with this not perfectly stigmatic light source, most of the scientific and technical obstacles have been identified so that additive manufacturing can find profitable application paths: layer-by-layer printing, the concept of voxel, this basic unit which allows the transition from a digital file to a 3D part, the use of supports, the importance of the viscosity of resins, etc. Incremental research therefore followed, along with more disruptive work leading to several other 3D technical processes (André 2017a, 2017b, 2017c). The disruptive intrusion of 3D printing in 1984 resulted in more than 180,000 publications, a development rate (publications, patents, trade) of 20%/year (André 2017a and references therein) because this technology has many advantages: the complexity is "free", it is possible to reduce the number of business interfaces between digital design and production, etc. (Cinquin et al. 2021). On this basis, the current market is around €30 billion/year and additive manufacturing has become an integral part of the concept of "Industry 4.0" (André 2019 and references therein).

Since the dawn of humanity, people have been making discoveries without the need for science. These are the inventors who create physical proofs of their ideas in their own environment. Today, a significant proportion of industrial innovations in high-tech sectors is based on recent academic research, even if the invention itself does not come from academia (Mansfield 1998). This is indeed what happened, but the concepts of photopolymer resins, lasers, digital technology and automation already existed. And anyone with a little education and the right knowledge could have come up with such a simple solution (see Figure I.2, adopted from André and Corbel (1994) and reproduced in André (2015, 2016, 2017a)). It probably took a somewhat *divergent*, even rebellious mind to step out of its academic research and be creative for a little while. This is a game between the self and the self!

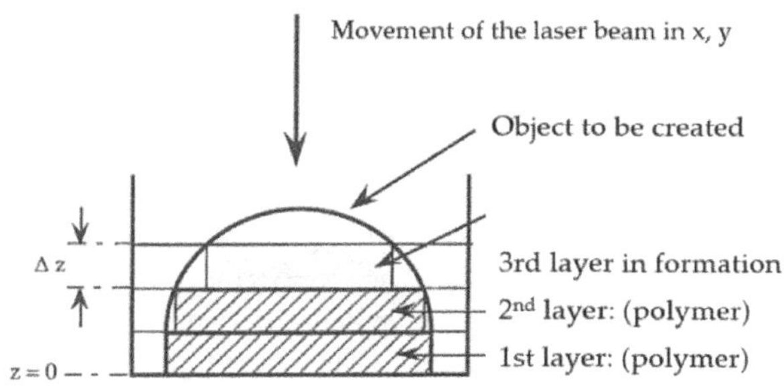

Figure I.2. *Historical diagram of the principle object layer by layer*

Before science was structured, there were still inventions and innovations following them; this was the case for agriculture around the year 1000, the iron and steel industry and the steam engine, etc. Everyone knows “the old gag” that science owes more to the steam engine than the reverse. This half-truth is often repeated because it makes you think. It was the interaction between the two domains that mattered because they reinforced each other (Mokyr 2017). It was Carnot who, 70 years after practical application, scientifically enabled progress in the operation of steam engines (Dubois and Brault 2021).

But it is quite another thing to bring far-reaching scientific knowledge to bear of mechanical energy that is profitable and useful to society, and its production in large quantities is another matter altogether. It is the missing link between invention-scientific knowledge (upstream or downstream) and innovation. And to achieve this goal, many people and professions had to be involved.

Unfortunately, CILAS managers, followed by a research organization associated with the first patent on additive manufacturing did not consider the technology interesting enough to maintain the patent (paternity of the process is now attributed to Chuck Hull and his US-based company 3D Systems). This shows the gulf that existed, and probably still exists to some extent, between boards of directors supporting the short term (the question of *decision*) and the world of research, which should support creativity, proof-of-concept the medium term. They failed to detect the potential of this French invention and bring it to fruition, no doubt because they were more concerned with a quick return on investment. We certainly need to be able to build bridges between the field and fundamental research in order to make more lasting commitments to transforming the imaginary into tangible realities. This means that we can take risks to support *creativity*. In Europe, certain research organizations, such as the EPFL in Lausanne and the Université catholique de Louvain in Belgium, have this pioneering spirit, without neglecting the fundamental. They succeed in bringing their discoveries to the field. France should, despite the political, make efforts in this essential area by enabling research to value *risk-taking* and (finally) put more trust in its researchers.

Today, there is a tendency to see additive manufacturing as a discipline, as something fixed from a paradigm point of view. But on the contrary, you still need to be undisciplined to innovate in 3D printing (but not only). But the *convergence* of several scientific disciplines: materials, processes, systems, design, human and social sciences, etc., remains necessary. *Interdisciplinarity* is not just a word, and it characterizes a *complex* but indispensable shared work process. Without it, it is impossible to test "good ideas".

Incidentally, we have succeeded – somewhat ironically – in abolishing layer-by-layer printing (ultimately going back to Battelle's old work) with colleagues from Fresnel (CNRS – Marseille). If we had succeeded in following this organization's patents in 1984, avoiding the use of layers, perhaps the future of 3D printing could have been different... I am currently trying to develop a process capable of printing a part in a single step, like a camera (or André Franquin's drawing in his comic strip on "small formats" over 50 years ago). It would be a radical evolution of stereo-lithography, but I cannot say much more about it at the moment – it is not patented, as I am having trouble finding the right light sources to produce a nice proof of concept.

Box I.1. *How can we believe that hierarchy follows your creativity?*

It is therefore necessary to make this relationship as fluid as possible for technological progress. But, for example, Ridley (2020) considers that regulation can harm innovation by favoring already-accepted innovations over new ones: between blocking competition and the public's desirability of invention, new players come into play: innovation is no longer technical, but takes on an increasingly social character (Chicago World's Fair in 1933 is therefore dead!), unless regulation blocks it.

According to Angeler et al. (2016), *disruptive inventions* are generally the result of the materialization of creative ideas:

> The fruit of an inductive process that sometimes spontaneously and unexpectedly generates a link between thoughts and/or objects that didn't exist before. Creativity is the cornerstone of scientific thinking, but academic scientists are judged by quantification measures that often leave little room for creative thinking. In many scientific fields, reductionist approaches are rewarded and new ideas are viewed with skepticism [the behavior CILAS management's attitude to the patent concerning additive manufacturing...]. As a result, scientific research is often confined to narrow but secure disciplinary ivory towers, preventing deeply creative explorations that could bring unexpected benefits!

As a reminder, science aims to discover, understand and develop universal theories, while technology is all about inventing through *creative, incremental and radical* processes of construction/deconstruction (Baumol 2002).

Maslow (1966) wrote: "A hammer tends to see all problems in the form of a nail!". Creativity, which is often put forward with its many clichés or comic images, is a necessary starting point. But when it comes to innovation, there is a *teleological* approach that forces us to think differently within increasingly complex and uncertain systems. This situation, relatively stabilized or rather controlled in the 3D field, has become more urgent in the 4D field (3D + adaptive matter) or bio-printing (4D printing of living matter), without being new in substance. It should induce a new phenomenon in research structures, particularly in academia, that of risk-taking in the event of failure (but in the event of success, whoever gets there first can win it all, especially if they are already recognized by their peers and hierarchy, etc.), which leads to a certain conformism or passivity.

On this basis of scientific knowledge, using the following definition of innovation: "the process of translating an idea or invention into a good or service that creates value with customers willing to pay for it", several application deployment activities that would give societal "consistency" to these new, highly open domains become necessary to move forward and invent (in a generic way, but more in a targeted way on specific objects).

In 2013, S. Tibbits of MIT wrote:

> An unprecedented revolution is underway. It is the ability to program biological and physical materials to change shape, change properties [...]. The idea behind 4D printing is to perform 3D printing with multiple materials, in order to be able to put several materials in it, and to add a new capacity to it: transformation. As soon as they come out, the pieces can transform from one shape to another, directly, all by themselves. It's like robotics without wires or motors. You print this piece, and it can turn into something else (Tibbits 2013, 2014).

4D printing was thus born by having a 3D printing of materials stimulated by an energy input. Figure I.3 highlights a very large number of degrees of freedom for 4D objects: three space parameters (classic 3D) associated with a temporal parameter, but also the nature of the active material and its energy excitation, localized or No. In the case of bio-printing (BP), it is nature that regulates the temporal evolution of the bio-constructed with cells that can differentiate and which are expected to approach a tissue, or even an organ. Figure I.3 (Demoly and André 2022) presents the general principle using a classic additive manufacturing process: first of all, it is necessary to be able to build (printability) an object by one known 3D processes with adaptive materials or adaptive renderings, and in a second phase, being able to change its shape from one or more specific stimuli (complex material-process coupling). This principle naturally leads to the production of relatively flexible (or locally flexible) parts, which must be able to deform or change functionality in the presence of a stress, on the one hand, of homogeneous or heterogeneous parts manufactured, if possible, using "classic" 3D printers, on the other hand.

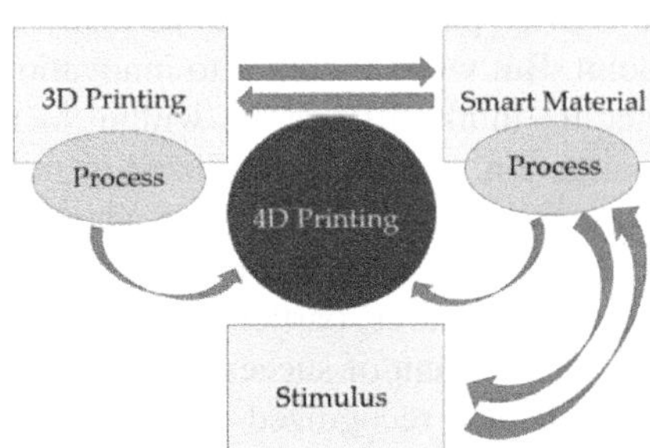

Figure I.3. *Process-material coupling, energy stimulation in 4D printing. For a color version of this figure, see www.iste.co.uk/andre/knowledge1.zip*

The situation with bio-printing is, as we shall show (see Volume 2 of the book), rather different, since the cells have to be supported using scaffolds, allow growth and differentiation growth and differentiation, etc. But in the end, the difficult control explained here becomes complex in the latter field, because cell fate is itself complex.

Box I.2. *4D printing*

In terms of invention, the 4D and BP conceptions should be seen as starting from several key words: *initiative, creativity, responsibility, trust, meaning, sharing, agility, process, flexibility, culture, heuristics, learning,* etc., these domains contribute to adding complexity to the issues raised in terms of 4D or BP object realization. 4D or BP manufacturing generates a technological inseparability superior to 3D printing, by producing an active object from the printer that must have an active future, rather than by activities organized in successive stages. There are therefore a number of information additions to be made and complex interdependencies. In any case, common lines of action should/can lead to a transfer of knowledge and know-how (Leleu-Merviel and Boulekbache 2014) as:

– Can we restore the idea, and its treatment due to its complexity, through a synthetic form of utilitarian formulation?

– How do we conceptualize this formulation? And how do we represent what we have conceptualized and/or designed?

– Do the tools of this interdisciplinary project representation influence the functioning of the project group involved in the realization of complex 4D or BP objects?

– What is the effect of incremental or disruptive design representations on exchanges between players: reductionism, standardization?

– Which tools can be used for a better understanding of design intentions?

– Definition of a model of design activity? And if so, how?

In this introduction, apart from the key words that punctuate the document (written in italics), we see a link between *invention*, which arises from a "one-off" idea, and *innovation*, which is basically the most general possible applicative result of the latter. On this basis, we remain true to the "old" leitmotiv of the Chicago World's Fair in 1933: science invents, industry produces and the public buys. This model has been much studied and much criticized in recent decades (Godin 2006; Balconi 2010). Satisficing solutions (Simon 1976), to describe those with which we are satisfied, for want of being able in practice to determine an optimal solution, corresponds more or less, for the moment, to what research can propose for 4D printing and bio-printing (BP). These are acceptable solutions via a "heuristic search", a trial and error selective search using acquired knowledge (hence certain conservatisms), but without the context having been finally associated with the operation. Figure I.4 illustrates the current situation (on which the scientific obstacles and the objectives particularly highlight the interdisciplinary, innovative, disruptive and/or exploratory nature of the projects).

Over the last century or so, the number of researchers in Europe has risen from a few thousand to several million, and this trend is continuing, despite a few disturbances. On the one hand, research activities have been the subject of a reassuring discourse on the independence of the researcher and, on the other hand, of a certain programming of research with a view to achieving certain objectives: security (before the fall of the "Iron Curtain", for example), economic development (from mass production with the ECSC (European Coal and Steel Community) projects to information and communication sciences and technologies), all of which contributed to various forms of competitiveness for France and the European Union.

On this basis, the stereotypical image of the scientist as the guarantor of truth and goodness is still part of the epinal image, which often positions him very high in relation to a social reality that he only imperfectly masters. The desire to achieve maximum "research efficiency" has led to the promotion of rather mono-disciplinary approaches that are easier to manage with "peers", the referents of a discipline. On the one hand, in-depth scientific research is supported by players in the same field, provided that the guarantee of excellence is defined and respected. On the other hand, it is easier for the state to make international comparisons on a discipline-by-discipline basis.

Indeed, and it is essential to remember this, we have gone, without realizing it too much, from a global scientific elite in limited numbers to mass research (with several tens of thousands of scientific journals), which represents a characteristic not discussed by all of the developed nations: research must, in fact, enable society to respond to the major challenges that arise today: jobs, progress, security, global warming, health and quality of life, sustainable development, etc.

Box I.3. *Developments in scientific research*

Independently of the criteria of spectrums of competence (scientific *paradigms*), which should not be disconnected from those of the societal world, several questions emerge in this linear approach as follows:

– How can scientific players from different backgrounds be mobilized to produce a proof of concept?

– Who pays for the cost of achieving proof of concept (or prematurity)?

– When taking out a *patent*, who owns the patent? How much leeway do the holders of the initial idea have?

– In the event of success, how do you get through the "valley of death" between "academic" proof-of-concept (POC) and the first step towards innovation with a "societal" proof-of-concept that respects the technological paradigm?

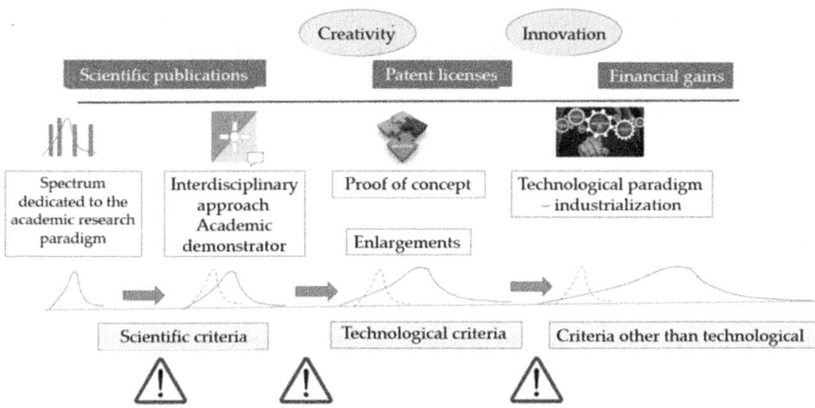

Figure I.4. *Stages reached (or to be reached) in the valorization process of 4D or BP technological research (excluding the innovation part). For a color version of this figure, see www.iste.co.uk/andre/knowledge1.zip*

The transition from "academic" POC versus "societal" POC is critical and must satisfy socioeconomic issues. In any case, it is necessary for "inventors" from the "hard sciences" to move away from the "artisanal" aspects they may have practiced (and which, naturally, may have been useful) to try to convince a wider community, acting on other criteria, and to steer their activities in directions of social utility by pointing them in directions/purposes that the players in these disciplines have not thought of. This involves learning how to achieve more optimal co-construction, as defined in Figure I.5.

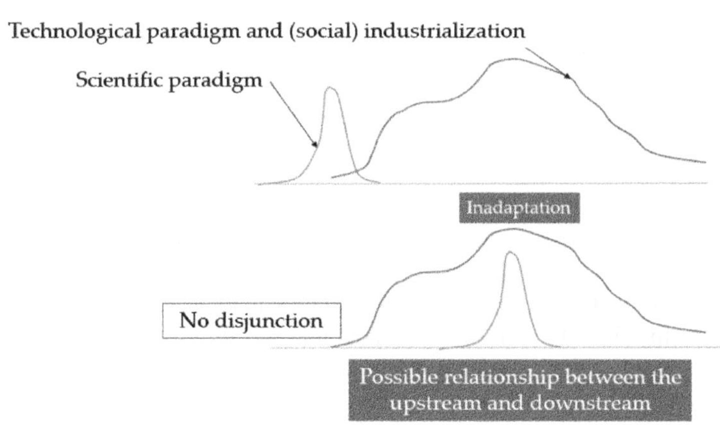

Figure I.5. *Moving (for example) from 4D to BP from "outside" the need to closer to it. For a color version of this figure, see www.iste.co.uk/andre/knowledge1.zip*

"The old scientific ideal of episteme, the ideal of absolutely certain and demonstrable knowledge, has turned out to be an idol" (Popper 1973). By bringing together the "hard" sciences and the human and social sciences (HSS), we need to master the links that exist between actionable knowledge and general theories in management science (and relationships), and in particular to know how to make scientific results more actionable. Creating "actionable" knowledge, that is, knowledge that can be used for action, making theories more actionable too, and bridging the gap between theory and practice are just some of the problem formulations that need to be part of HSS know-how (see Figure I.6). If we are talking about 4D printing or BP, this means being able to propose "well-defined and transferable processes designed to produce certain results deemed useful" (Friedman 1952). In this context, it is essential to define clear rules for operating modes in relation to an applicative purpose, and thus to translate the possible complexities mentioned above in terms of usable reductions.

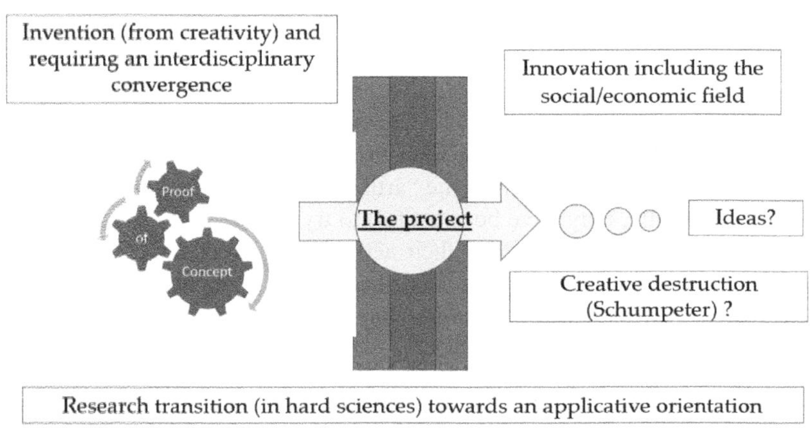

Figure I.6. *An interactive project for the invention–innovation with feedback. For a color version of this figure, see www.iste.co.uk/andre/knowledge1.zip*

The research methods are of the collaborative type, in partnership with organizations that are pioneers or wish to experiment with new ways of organizing themselves by working in the field, with partner companies and organizations, to better formulate and solve innovation and change management issues. For example, today, capturing the attention of consumers and meeting their needs through digital tools is a priority. With and after the crisis linked to Covid-19, it was easily conceivable that the year 2022, via the interoperability of players, marked the emergence and development of integration applications that could be adapt to a number of technologies, thus allowing a gain in efficiency and responsiveness for companies.

Regardless of this aspect, the idea is to advance management science research, but also to produce applied knowledge (concepts, models, tools, methods and analysis grids). Emerging 4D/BP issues are good "candidates" for this type of work. This type of work has two components, one aiming for a science-technology-society resonance, in which 4D/PB is seen as an applicative goal, extending developments in additive manufacturing, and the other where 4D/BP is merely a means of exploiting the potential applications of intelligent materials (see Figure I.7).

This "supply" activity should be confronted with knowledge of what is real and what is possible using multi-criteria decision support. Research into decision-support tools lies at the intersection of several disciplines: psychology, sociology, management sciences, economics, law, but also scientific and technological knowledge, with a particular focus on the sciences of management innovation and technology management.

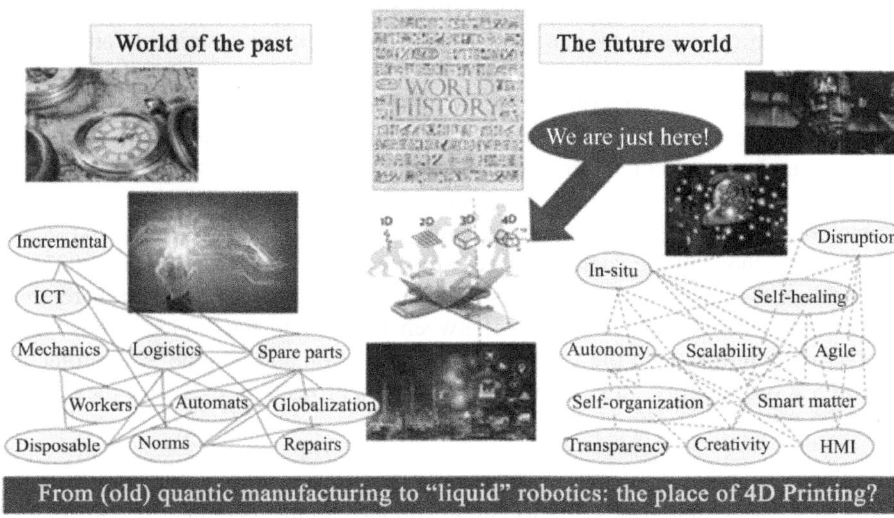

Figure I.7. *For a future use of "smart" materials? (ICT for information and communication technologies; HMI: human–machine interaction). For a color version of this figure, see www.iste.co.uk/andre/knowledge1.zip*

The aim is to understand how we move from invention to innovation. According to Bourdieu (1997), it is a matter of mastering the communication between science and economics: "the stakes are not the same, the ends are not the same, the agents have completely different, even opposing, philosophies of existence, and this generates profound misunderstandings". And yet, we cannot talk about innovation without returning to Schumpeter (1934, 1947), an essential figure whose work serves as the backbone of the field. To put it simply, three levers of growth must be taken into consideration:

– the existence and development of ideas (creativity, divergence, etc.);

– support for innovation (even if only through patents, technology *transfer* and *technology transfer* to avoid yesterday's innovators turning into conservative rentiers);

– creative destruction (the new replaces the old).

In order to achieve acceptable growth, government support is required in a number of areas (including research and training), as well as creative researchers and entrepreneurs (with at least a pioneering spirit), backed by an enlightened, if not complicit, tax administration.

Then, the power of creative destruction identified by Schumpeter, and that of the transformation of capitalism, can steer society toward more sustainable and shared technical prosperity. The concept of creative destruction is a process which implies that new innovations, rendering certain technologies obsolete, enable the creation companies to compete with those already in place. So, overall, new jobs, new activities and new products are created, replacing those that already existed. In France, however, the "Sapiens" think tank points to a scientific backward trend, characterized by a decline in master's enrolments in the hard sciences and health (-1%), while those in the social sciences are soaring (+10%) (Maddyness 2022). In this reference, A. Rossmann, an industrialist "exiled" in the United States, reminds us that "one of the major differences between France and the United States lies in the fear of failure [...]. In the USA, we remove friction and assume 90% protection to gain 5% profitability. In France, we want 100% protection, even if it means only 1% profitability".

On the one hand, technical progress generates an improvement in the financial health of states, which makes it possible to support the development of fundamental research, which itself should (in principle) support invention through its creativity and, consequently, innovation. But, as Oakey (2015) points out, independently of the measurement of the number of invention patents, there are no precise data concerning the overall impact of this inventive research on the economic growth of countries, because the performance of companies and that of the technologies they use and/or produce are difficult to dissociate from other effects. In these circumstances, economists have often resorted to measuring the quality of invention and innovation in terms of affordable input costs. Moreover, for Apolitical (2019), financial pressure can be favorable to innovation because it supports change, but too many resources can make researchers "lazy" and can encourage them to rely on their laurels or rather to develop research, which will not be rapidly accomplished.

On the other hand, this reductive description also includes the buyer, who may or may not accept a new product. They therefore have an eminent place in the future of an invention and, increasingly, they are becoming true partners in the design of new products. Co-creation initiatives, which involve civil society, help us to democratize innovation processes in several ways. By including beneficiaries from the outset, they can help steer innovation efforts in the most relevant and appropriate directions. Future adoption by users is facilitated, while the legitimacy of the innovation effort is strengthened. Finally, raising awareness and communicating more widely to civil society about the important challenges and the role of innovation in encountering them are another advantage of co-creation initiatives (OECD 2021).

On the other hand, innovation is everywhere, in this world of material goods first and foremost, linked to preliminary knowledge of increasingly scientific origin, but also in other fields, particularly cultural and stemming from the human and social sciences. Innovation can be found in the popular imagination, in fictional literature, in the media, normally in public policy, and is part of everyone's vocabulary. The word "innovation" is associated with the notion of modern society, "a panacea for many problems and a phenomenon to be studied" (Gaudin 2008). Helga Nowotny defines our times as a fascination with, and quest for, innovation (Nowotny 2008, 2006).

According to Accenture (2019), the need to innovate to ensure the competitiveness and viability is a notion understood by the vast majority of business leaders and employees who wish to participate in innovation. Some of them claim to be supported by their superiors. From an organizational point of view, the question arises of establishing a culture of equality, supported by a culture of diversity to increase the level of innovation in organizations. According to this report, it is the combined effect of equality and diversity that should be sought for a more innovative work environment. This social aspect completes what has already been presented.

Innovation is therefore, unlike invention, an activity of *establishing relationships*. Technology identification is not directly linked to invention, but to successful invention. It requires, within the current determinism, the identification of a need or potential market for a new product or process (Freeman 1988). In addition, it relies on technical knowledge that is, if possible, already available, but with new contributions from scientific and/or technological sources. In general, these are improvements resulting from incremental research. These improvements can be small or large, and apply to all technologies, old or new. Disruptive innovations, such as those resulting from the invention of the transistor in the last century, the fruit of initial scientific research, significantly change the relationship between producers (new products, new training, new machines) and users. The market is disrupted.

This has been the case for several years now, with the development of digital technologies, artificial intelligence, robotics, home automation and additive manufacturing technologies (to stay within the technical domain). Under these conditions, the relationship between technology and society evolves over time in a nonlinear fashion, with each wave of major inventions reshuffling the deck (with, in international competition, the promotion of the idea of "invention of inventions" (Landes 2006)). With disruptions linked to a "local" invention, the realization of new combinations of innovations as they trickle down, known as "systemic innovations", can be envisaged, leading to an even more global disruption (Krichen and Chabaud 2019).

Irrespective of this starting point, it is necessary to move from an "academic" proof of concept to pre-industrial prototypes in order to envisage the design and development of an industrializable project, with the launch of production, marketing and communication. During these various stages, technical improvements and the market are constantly adjusted. But this process remains uncertain, indeterminate and trial-and-error, and there have been some spectacular failures with good ideas (ideas "too far in advance" and resistance to change, etc.). Remember that the gap between technological invention and market acceptance is known as the "valley of death" (Markham 2002; Zhou and Wang 2020). We need to overcome some breaks with classic operational research (Taylor 2008) in management limitations of *decision models* and the difficulties associated with decision optimization. In doing so, a paradigm is defined by three pillars: consideration of multiple criteria to aid decision preference for reasonable and robust solutions rather than strictly rational ones, and a conception of the helping relationship in decision support based on the search for coherence with values rather than the search for an optimum. In a second phase, this work should reveal operational trends in the two interdependent areas of 4D printing and intelligent materials, and bioprinting and cellular material, in the medium and long term.

Several research questions will thus be considered to deal with the propositions/prescriptions of change, such as:

– The existence of creative people, *imaginative* but *responsible* in their description of their inventions ("the rejection by scientific thought of all considerations based on notions of *value*, perfection, harmony, meaning or end and finally the complete devaluation of being, the total divorce between the world of values and the world of facts" (Koyré 1988)).

– The optimal nature of the (lay) *co-creation* group(s) (Naudon 2013; Deserti et al. 2022).

– The necessary steps in the co-creation process within a framework of social responsibility (Wittrock et al. 2021) based on the construction of a mental representation at the interface of disciplines (Bourdin 2022).

– The construction of decision-making processes of group(s) involving actors with different roles, skills, knowledge and expertise (according to Jakobsen et al. 2019), conceptualize multidimensional approaches linking the scientific, material, discursive and organizational dimensions of innovation and their synergistic effects on major societal challenges).

– The search for the representation mode(s) of a 4D/BP system and the diversity points of view.

– The key role of public representatives in the various stages of the co-creation process 4D/BP.

From a practical point of view, co-creations make it possible to master the way in which the dynamics of multidimensional and multi-scale innovation can be initiated and followed to envisage the creation of economic value (with parallel solutions participating in environmental and social challenges). So, it seems advantageous to test the proposals to guarantee the relevance of the final product, requiring feedback to refine the technical aspects of the innovation. To address credibility issues, transparency in the process, including prerecording assumptions and filing of an action plan before execution, should eliminate the temptation to report the best results after completion of the project or to revise the assumptions once the results of the study are known (Tsui 2016).

The notion of Living-Lab or living laboratory (Bulkeley et al. 2016) is understood as sites designed to design, test and learn innovation processes in real time, reducing the uncertainties associated with the new. These structures envisage responding to particular societal, economic and environmental challenges. By providing a space for negotiating problem definitions and understandings, diverse resource claims, authority or dominant ideologies, the living lab has the potential to contribute to changes in innovation practices.

In a situation of stability, where the incremental reigns supreme, results are accumulated and publications are subject to known regulatory mechanisms, which normally allow us to envisage improvements to marketed processes. But when these are no longer functional, or when new processes are invented, disruption through divergent revisiting imposes itself as a *disruptive* element in a possible way out of impasses or simply astonishments. Thus, as Edgar Morin (quoted by Bouiss 2020) explains: “crisis depends on randomness: at certain crossroads, it is possible for a minority, an individual action, to tip development in a direction that is sometimes highly improbable”.

As such, invention and innovation are cultural phenomena associated with the economy and finance, translated into products or technical methods. Future research on invention and innovation could therefore be strengthened by integrating theories of meaning into their explanations.

I.1. Between invention, innovation and 3D/4D/BP printing?

> The acceleration of life's rhythms made 19th century man realize that he was immersed in an absolute flux, where every stable entity revealed a fiction, fragile and provisional. This revelation brutally destroyed all the reassuring support of beliefs and reference points that ancient societies had built up for themselves (Stiegler 2021).
>
> Science cannot think (Heidegger 1968).
>
> Strategic intelligence is the ability to think in the world, starting from the world. French intelligence thinks the world from the inside out (Saint-Étienne 2015).
>
> Evolution, then, is in all cases a change from a more diffused or incoherent form, to a more consolidated or coherent form (Spencer 2017).

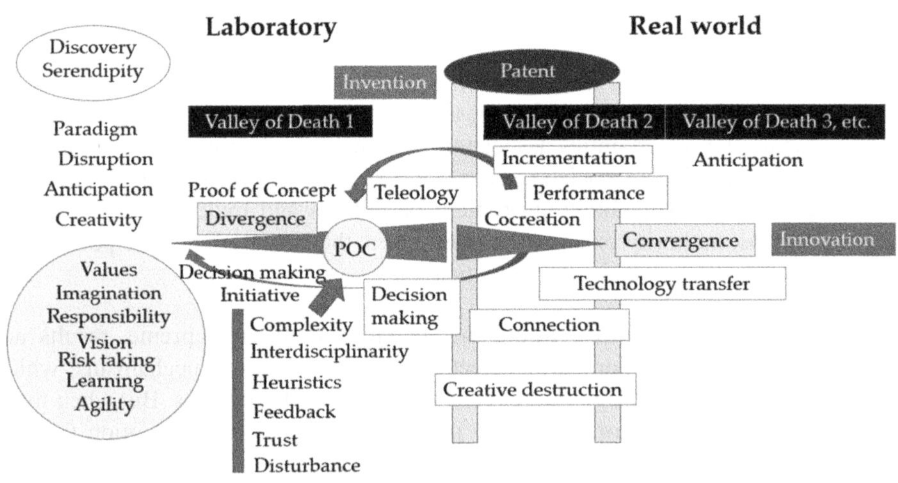

Figure I.8. *Keywords involved in the relationship between invention and innovation. For a color version of this figure, see www.iste.co.uk/andre/knowledge1.zip*

In the body of this introduction, the author has highlighted key words; these are shown in Figure I.8 and correspond to elements that are a little more in-depth than their simple introduction in the text. While these general questions did not have to be considered in the author's personal case of 3D printing in the 1980s, because the patent was placed in the public domain in 1986, these various aspects of connection, creativity, invention, innovation, in short, the coupling between creative science and society, need to be mastered, if not known in their general aspects. This context forms Chapters 1 and 2 of Volume 1 and Chapter 1 of Volume 2 of this book (see Figure I.8).

In the second part (Chapters 2 and 3 of Volume 2), the three areas just introduced will serve as examples of successful, ongoing or failed valorizations. It is obvious that everyone is familiar with "3D printers", and this technology which is growing steadily of around 20% over the past 20 years. Independently of the initial French failure recounted in this introduction, the author felt it useful to examine why (and how) this highly visual technology was able to develop when other relevant subtractive (material-removing) technologies exist. The 4D and BP situations, much more recently (around 2013), correspond to immense needs in the medium and long term with considerable market potential (which has nothing to do with the global 3D business, which is on the order of €30 billion/year (André 2017a)). And yet(!), the reasons for these stumbling blocks, of various origins, will be analyzed and compared with additive manufacturing (even if it is less than 40 years old!).

These analyses and comparisons serve a personal purpose, corresponding to the enhancement research work on additive manufacturing and, more recently, 4D printing and, to a lesser extent, bioprinting. Trained many years ago in an engineering school with an essentially mono-disciplinary cultural background, innovation for the author was rather centered on the aforementioned profession of faith from the 1933 Chicago World's Fair.

But personal experience shows that things are not quite so simple, and linearity is a utopian dream. It was therefore important for the author to learn, for example, by reading literature, which provided a (no doubt partial) outlet for the poor researcher lost in the meanders of valorization (and it has to be said that, even in what some call the startup-nation, they are not helping!). This book serves as a kind of testimonial and a place for sharing!

These analyses also help us to understand how the future of nations engaged in international competition is played out: while perverse effects (inequality, job destruction, social mobility, etc.) exist (and are growing), the capitalist system, which has enabled high levels of prosperity for around two centuries (amplified by a factor of around 100), remains a good option for many citizens, as long as creative destruction is likely to create (local) growth (with appropriate financing upstream research, the knowledge economy and support mechanisms for innovative companies). The current constraints of sustainable development, depletion of reserves and alternative energies represent a real challenge for Schumpeter's successors.

So, locally, companies are called upon to maintain or even develop their performance, if they do not want jobs to be relocated. The dynamic and *anticipation* required to stay in the race and maintain a position that must remain "acceptable", with purchasing power maintained or increased (if only by being able to offer the training that will be needed tomorrow).

In the absence of integration of society (including society at work), social inequalities can, according to Case and Deaton (2021), become a scourge in terms of health (and possibly of the future). So, with ever-lower wages for the poorly educated working population, the weakness of collectives (such as unions and other counter-powers), the world of private university education in the United States is no longer accessible to the children of the latter (prohibitive tuition fees relative to wages), leading to a spiral not very conducive to alleviating the problems briefly mentioned here. There are many forces that can act directly or indirectly, but the obvious one is that the "left-behind" is not involved in the economic power of, for example, the NYSE (New York Stock Exchange), nor are they adequately connected to the functioning and use of the healthcare system. Case and Deaton (2021) attribute to the destruction of the white working collar class the failure of the U.S. healthcare system on the one hand, and of a model of financial restitution too oriented toward the poor on the other. One of the reasons cited is the relocation of industrial activities to China.

Against this somewhat anxious backdrop, negative trends can accompany growth. Aghion et al. (2020) and Case and Deaton (2021) show how some countries are better than others at avoiding them (for France, for example, the MESRI (2022) note could be used as a basis for further reflection). In particular, according to their analyses, younger companies enjoy net growth relative to older structures; similarly, growth in GDP/per capita increases in countries where creative destruction (which is more or less the same thing). What role will 3D/4D printing and BP play in this setting, where the risk of production (at least of machines) can, as in 1984, take place outside our country?

One possible response may be linked to France's deindustrialization, revealed in part by the Covid-19 pandemic. For this country, which is at the bottom of the rich world in terms of its manufacturing component, we need to act as Vicard and Bensidoun (2022) point out:

> However, this does not prevent us from identifying the sectors on which we agree to be dependent (and on which partners) and those on which we would like others to be dependent on us, so as to create real interdependencies preserving the countries' autonomy of action and gains from trade.

This forward-looking approach must include collective choices on technologies and "sectors of importance to the national economy and the comparative advantages of the French economy and French companies in relation to our partners" (Vicard and Bensidoun 2022). It is all about leading the competition.

According to this note from the MESRI, in 2019, France's research effort (GERD/GDP ratio) was 2.19%, including 1.44% for the research effort of companies. For Sweden, Germany, Austria and Belgium, this ratio is above 3% (not to mention countries like Israel and South Korea with ratios above 5%). This weakness, which is not temporary, can therefore have an effect on the performance of companies and on the economic health of our country. Moreover, the High Commissioner for Planning recommends a national strategy for the reconquest of the productive sectors, agriculture, industry and services, supported by a large public and private investment capacity (Bayrou 2022). He believes that, unlike the financial crisis of 2008, the current crisis reveals a systemic change in our economy which gives France all its chance for a real growth shock, accompanied by the appropriation of the triptych "debt-production-instruction". He writes: "France has a deficit of some sixty to seventy billion in its foreign trade when Germany comes out with a surplus of 250 billion. On this criterion, in 2000 we were ahead of Germany, and the dropout materialized in twenty years".

When it comes to figures, according to Verdier-Molinié (2022), France's financial situation isn't exactly stellar:

– 7 billion fewer hours worked than the European industrial countries of Belgium, Germany, Spain, Italy, the Netherlands, Austria, Sweden and the United Kingdom;

– 2,950 billion in public debt, despite "the highest compulsory taxes in the world", increasing debt from €20,000 per capita in 2012 to €44,000 today;

– 1,120 billion in compulsory levies, or 47.5% of GDP 2019, when our competitors are on average 5 points lower;

– 1,460 billion in public spending (55.1% of GDP in 2017 to 59.9% in 2021);

– more than 2 million jobs lost in industry since 1980, when the country had 5.5 million industrial jobs. By 2021, this figure will have fallen to 3.1 million, reducing value added in industry to 280 billion euros, compared with 300 billion in Italy and 752 billion in Germany;

– 82 billion trade deficit, while Germany posts a positive 182 billion (250 billion for Bayrou 2022).

Can innovation undoubtedly be a positive element in winning back the market?

"We have a tendency to expect everything from the state, including billions for hospitals, schools, the justice system and the police [...]. Our conclusion is that this distribution of money solves nothing. It's not money that heals, invents software or produces bread. It's people" (Montenay 2022). In the report published under the responsibility by Guillaume Kasbarian (2019), simplifying administrative procedures is one of the main ways to improve our country's competitiveness. It highlights the major obstacles to reindustrialization:

– legal uncertainty;

– the opacity of procedures, the steps to be taken and the lack of predictability when it comes to deadlines;

– sometimes excessively long processing times;

– pilot failure;

– the absence of a shared culture between the administration and project developers;

– a lack of support in the face of environmental requirements that are a source of complexity for manufacturers.

For IREF-Europe (2022),

> The first observation is quite edifying, but also bears witness to the blindness of our politicians. They want jobs and growth, but don't seem to measure the cost to business of the standards they are constantly creating. One law of this five-year term epitomizes the normative madness of our politicians: the 2021 Climate and Resilience Law. A 126-page text packed with standards and regulatory constraints. Perhaps that's why manufacturing's share of France's GDP has fallen to 10.5%, even if eurozone ministers believe that (just) the lack of European funding is slowing the growth of startups and the tech ecosystem across Europe, but also of the European economy as a whole (MEF 2022).

But if we follow Valéry (1957), the path will be difficult. He writes: "the reason people don't like darkness is because it forces them to think, and they don't want to do that, because it's too hard. It's much easier to accept things".

Box I.4. *A quick look at France's socioeconomic situation*

In the light of the literature and Aghion et al. (2020), it is possible that there is a common thread between the core of this book and the disruptive inventiveness of academic research. Creativity in research generally takes place within a framework of indiscipline, requiring an examination of sensitivity to otherness, the unexpected, the new, disruption, in short, the creation opportunities, that is, "revolutionary" opportunities. This is what the good old scientific culture full of certainties does not necessarily like (because, historically, it's been tried and tested), those that can (must?) be taught (and, to be honest, those that do not fit into the reductive drawers of official science, because they are essentially based on regularities that only allow for incrementalism).

In its working document (COP 2019) between MESRI and CNRS, it was written:

> In this way, disruptive innovations are essential to conquer new markets, or simply to keep them, and to create or maintain jobs in France and create value for our society. Last but not least, the contribution of science is vital to achieving the goals of sustainable development, particularly with regard to global warming, renewable energies and health, but also to meeting other major challenges facing our societies, such as such as security, AI and migration.

This view is supported by Christian Lindner, who stated in 2022 (MEF 2022):

> Start-ups and scale-ups are essential for innovation, growth and job creation. Together with our European partners, we will use public funds to attract more private capital into this sector, particularly at the startup maturity stage, and thus strengthen the innovation potential of the European venture capital ecosystem. Improving conditions and providing the right incentives in this area will boost the economy as a whole. For it is now that we need to lay the foundations for a digital, sustainable transition from a crisis mode to a focus on growth.

> The practical value of all programs and proposals based on such a belief is almost nil, but the damage they cause is immense, even if they remain only on paper, because they divert people's attention from essential and urgent – if hardly visible – tasks and raise hopes whose non-fulfillment leads to the lethargy of despair (Röpke 2022).

All that is left is to create!

Box I.5. *Transitions to bounce back*

In this highly viscous environment, the undisciplined person who steps out of the shadows may be considered an obstacle to thinking in circles, or even a nuisance. By putting certainties up for debate, he can pose a threat, especially if he has a chance of bringing his project to fruition. At least from the outset, if the project is successful, it leaves the field of possibilities open and can continue its adventure.

The problem is the same whether we are talking about disciplinary or interdisciplinary (it is not uncommon for a new discipline born of the "fusion" of several fields of research to behave in ways that are far removed from the freedom and tolerance it claimed to embody at the outset; Schumpeter's theories would therefore be applicable to scientific research!).

However, public confidence is on the wane. We are living in an age of disenchantment, where the emphasis on the collective, on major projects far removed from the citizen, and the promises associated with them, can give rise to mistrust. But above all, any eloquent vision must be nourished by a solid memory steeped in previous experience, and the avoidance of contempt for the uninitiated; all resources whose appropriation by the powers that be is less and less shared. Just because the conceptual generality of the subject matter is coherent does not mean it can be of use to the average citizen. It is not a question of using some sort of rhetorical artifact, prefabricated for the circumstances and whose reductive rendering is akin to a formatted dissertation and, ultimately, not very realistic; you have to start from the ground up, from what is possible, from your dreams and your needs!

In this context, Silberzahn (2022) speaks of a candidate for the presidency of the French Republic who would like to create a Ministry of Innovation attached to the Ministry of Industry to redeploy industrial activities. He is even followed by a scientist, Laurence Devillers, who would like to see a Ministry of Digital Innovation (Devillers 2022). This sounds like a logical proposal, because the situation is serious and deserves major efforts. If we want to measure the limits of this false good idea, we need only consider the proposals of Japan's MITI Industry and Innovation Ministry:

> According to them, Japan's post-war economic miracle was due to its expert hand and vision [...]. Many experts criticized [the West] for its lack of vision, its lack of voluntarism, the chaos of numerous small initiatives when it was necessary to think big, and the myopia of the market when it was necessary to see far [...]. Japan was entering a recession from which it never really recovered. The famous [...] plan remained nothing more than a plan produced by technocrats who had come to believe their own verbiage. Since then, the country has missed out on just about every major revolution (Silberzahn 2022).

This model, according to which a lack of innovation can be dealt with from the top down, does not work, and all of the "grand plans" inaugurated by the authorities (named Calcul (Computers), Concorde (Aeronautics), etc.) have ended badly because politicians who take themselves for Colbert persist in thinking that solutions can only come from an elite capable of thinking the future. "In 1994, to take just one example, they were the ones who decided that the Internet was not designed to enable commercial transactions, at the very moment when Amazon and Netscape were being created". For the "grass-roots innovator", this situation, where money is lost in vain activities, is a real injustice, a "valley of death" (and there are plenty of them in this book!).

However, according to Harari (2018), human beings are distinguished from the animal kingdom by their ability to believe in fictions, fictions that must be reductive if they are to be understood by all. Although intersubjective, belief in a positive common goal can have consequences for the social groups that integrate them, giving them the power to maintain and even amplify the dynamics and cohesion of complex human communities. In this way, fictions can shape our world in a tangible way, but their practical application in the field may not...

In *Fiasco* (Grisoni 2022), it may also be useful to read about projects that have been put aside, forgotten, aborted, failed, etc. A process of "societal extinction" can be linked to the elimination of original processes that have not reached the stage of being purchased by the public: they disappear from the market and can quickly be forgotten. In this way, we would be collectively incapable of building a future that is properly sustainable/fiction without knowing about past mistakes (which we could reproduce); at the same time, more to the point, forgetting rejected processes does not allow for a satisfactory representation of internal or external causes. This phenomenon already exists in research, with the difficulty of publishing negative results. Herbet et al. (2021) remind us of the demand from the research community for the publication of such results, because they contribute to the construction of the scientific path, but despite this adherence, their dissemination through the traditional channels of scientific publishing remains very limited. This situation is particularly acute when results that do not respect the "official paradigm" run counter to the preexisting consensus (Hammond 2021).

Moreover, the public deficit is increasing, debt is growing, there have never been so many dissatisfied people, the job-salary situation is deteriorating, the trade deficit is widening, school performance is more and more mediocre, society becomes ossified; in short, nothing is going well in the "kingdom of France". More than a century ago, many pioneers were French, for example, in photography, cinema, automobiles, aviation, etc. But, obviously, the world has profoundly changed and if French pioneers can still exist, it is necessary to release, in a less stationary organizational framework, the living scientific forces to bring out meaningful innovations (and ultimately jobs), while drawing on our disciplinary skills. Creating value means knowing how to support successful ideas and take risks to lead them – when it makes sense – toward their industrial and/or economic achievement. This obvious consideration today highlights some key words such as those presented in Figure I.8.

But being isolated probably does not get you anywhere, and neither does getting too involved. This means that not only do we need a strong commitment to supporting innovation (this is a prerequisite and, according to The Innovation Research Interchange

(2022), is generally a given), but also a modest critical size to provide the necessary flexibility but sufficient to ensure that the results are part of the fiction proposed by the states, with the necessary dissemination to ensure recognition of its performance and a profit. Wilson (2022) proposes various methods to promote innovation:

– Crowdsourcing with internal outsourcing. It gives rise to new ideas and perspectives that can create value for companies.

– Innovation open freely shared between players in the value chain, which also offers companies more opportunities to create value.

– Innovation management software for measuring the progress of innovation. However, the analysis grid can be questioned.

– Invention comes from people who cooperate with the company and rely on the deepening of scientific knowledge (a characteristic of the most successful companies according to Wilson (2022)).

– Prototyping, which involves creating preliminary and test versions of products. This method reduces risk and increases a product's performance on the market.

The same author reminds us that design-thinking is often advocated, and points out that "phased" systems "are based on a linear process involving a series of sequential steps to launch new products. However, innovation processes are rarely linear, and a rigid formalism is not applicable to most industries. They may be best suited to the manufacturing, IT and healthcare sectors, but not in other areas". Design-thinking is an approach that uses a sensitivity parameter and the support of a designer to match consumer needs with what is technologically feasible. It is, for Wilson, "nebulous and even vague among those who practice it. [...] The ambiguity of design-thinking is why only companies in the arts and entertainment, retail and marketing sectors have felt its benefits".

For his part, Christensen (2021) proposes several ways to improve business competitiveness and stimulating innovation:

– *Diversity*: necessary element for innovation. Homogeneous teams with the same culture, particularly scientific, experience few conflicts, but a heterogeneous, interdisciplinary team, when well managed, can lead to decision-making which incites innovation (we will come back to this aspect in the following chapters of Volume 1). But, in the dynamics associated with technological progress exploiting diverse knowledge, brought to bear by equally diverse profiles, managers must ask themselves the question of the relevance, at any given time, of the polydispersity of the know-how of the company's partners. Otherwise, it is possible to gain in terms of performance at a given moment, performance that will be lost due to the unsuitability of the players in charge of innovation at another time. The initial competitive advantage will be lost.

– *Compatibility*: companies need to move beyond local specificities if they are to develop beyond borders and, ideally, worldwide. This implies a cultural and technical "compatibility mindset", that is, meeting the needs, tastes and desires of customers the world over. Christensen quotes Bjarklev, in Denmark, who wrote: "The key is to merge part of the world into your local scene". Indeed, this academic welcomes students from different backgrounds to create a common culture that is not entirely Danish by encouraging cross-cultural thinking. But what about the performance of certain specialties, exported as they are? Caviar, champagne, foie gras, etc.?

– *Proximity*: in this, contact with an eminent person can be at the origin of a particular creativity because conversing with them and trying to present their ideas to them while trying not to disappoint them is likely to bring out some creative elements. Christensen sees this description as a practical manifestation of an important part of Porter's cluster theory (1980, 1985) who described notions of "sustainable competitive advantages" that would arise "from more and more local things – knowledge, relationships, motivation": the proximity of a leader or a model can result in forms of innovation, creativity and problem solving, which can lead to a competitive advantage. What is the role, in the diamond made up of the following elements: the ability to produce locally, the nature of demand (local or global), the existence of complementary local activities, the company's strategy, clusters? They are defined as sets of suppliers, customers and related industries that are all competitive in their respective fields. It is with clusters that a country can make its mark on the global level, because the existence of a cluster provides fertile ground for innovation and, in so doing, for the creation of comparative advantages against competitors. This is why Porter advocates the specialization of both firms and economies. It is therefore better to bet on the creation of specialized sectors (competitive advantage) starting at the national level by taking its marks. Christensen's approach is moderately appropriate here. On this subject, the Minister of Innovation, Science and Industry, François-Philippe Champagne of Canada writes: "we want to strengthen the research ecosystem to make life sciences an industrial pillar of the Canadian economy". But, in Porter's logic, in the "virtuous circle", what is predicted is that the more pharmaceutical companies and industry suppliers there are in Canada, the more other players in the world will want to join this new life sciences cluster (Normand 2022).

– *Respect*: proximity and the absence of intellectual hierarchy for those involved in different disciplines in the innovation process no longer separate those at the top from everyone else. If this is a fact, the question is whether or not this disrupts the efficiency of the project group involved in the innovation. So, for this author, showing respect is a way of instilling a way of thinking that recognizes that the best ideas can emerge from anywhere, especially those who are close to customers. According to Esther Wojcicki (whom he quotes), respect can be a considerable as it can encourage people's aspirations by seeing them as truly capable, while strengthening their determination to persevere and succeed. It is, basically, a mode of trust that must be established within a heterogeneous group for which there is no quantitative mode of evaluation of individuals in relation to the issue concerned.

– *Curiosity*: the aim of The Global Curiosity Institute is to help companies use curiosity to innovate more effectively. Stefaan Van Hooydonk, the institute's founder, has defined curiosity as "the engine of adult success and [...] the gel that creates positive links between people" (Learning Technologies 2022). Curiosity, the desire to see if the grass is greener on the other side, is an element that helps stimulate higher order thinking to generate new ideas. "For companies that must continually evolve or die, curiosity is the saving grace" (Christensen 2021). In fact, curiosity and creativity go hand in hand. To express creativity, we can speak of intuition, attitudes, sensitivity, a divergent and playful mind and so on. Basically, it is about putting names to complex concepts, without hoping that this is enough to describe them. Creativity is not easily "caught", but some of its manifestations can be understood, and there may even be methods of creativity. To begin with, one method seems easy to implement: to make known phenomena unusual by questioning them, inverting them, transforming causes into consequences. In short, to enter into the bissociation dear to Koestler (1976). While these processes do not systematically lead to new ideas, questioning can lead to a renewal of ways of thinking and acting. This position may be held in particular by those who consider experts to be the least apt to discover a new idea, but there are others. But there are other conservatisms at work, and this assertion is not demonstrable. The next step is invention and its aftermath, all of which is the subject of this book.

However, these various elements, which are mainly drawn from the humanities and social sciences, are only imperfectly reflected in the selection criteria for proposals for financial support from the European Union, where the term TRL (technology readiness/maturity level), is translated by a TRL scale that evaluates the level of maturity of an activity aimed at technological applications. It starts from 1 for basic research and reaches 9 for applications.

The European Union's response to the needs of its citizens through its R&D framework programs is quite simple in terms of innovation: it is better to go incremental with TRL values, in the predictable, in continuity, with only marginal adaptations to keep the old principles that no longer work very well, but which are accepted by the countries of the Union (UE 2018). A plea for disruption is a strong strategic element on how the EU intends to take global leadership in certain key areas. After all, neither Amazon nor Apple was created in Europe, and the EU's strategic plan presented here could have been written by any slightly enlightened person who has watched a few in-depth TV programs on the future of the world. So, let us make a big deal of digital: while forecasts for 2050 show that 50% of the energy produced by that date will be consumed in the digital sector (André 2019), sustainable development, low-carbon economies, the circular economy, green deals, etc. This may not be useless, but it is probably not enough.

It should be remembered that digitization reinforces globalization by accentuating the movement toward fluidity, as pointed out by Hartmut Rosa (2010). He wrote:

> Acceleration goes hand in hand with petrification, change accompanies inertia, and contemporary society is perceived as both fixed and frenetic. But we can try to change this by moving from traditional programming to real time for what can be mobilized, from the short term to divergent ambitions for the future of Europe.

And, to do this, will we be able to count on the numerous expert committees linked to decision-makers of all stripes, who will leave the initiative to the researchers and who will accept the risk of failure in their attempts to create something new?

NOTE.– TRL:

The scientific and technological activity concerns rather incremental innovative work authorizing industrial developments in the short term (approximately 5 years). The object of the Union's plans is not to engage too much (apart from the European Research Council (2018)) in long-term risky operations with TRL values (see Figure I.9, taken from (Mapping-Consulting 2012) for the upper part and from (DGA 2009)). The expression TRL defines a degree of technological maturity (TRL). The TRLs form a scale for evaluating the degree of maturity reached by a technology. The TRL scale was adopted for the main purpose of managing technological risk in programs, with a few minor adaptations (replacement of the notion of space by the notion of operational environment) (DGA 2009).

The notion of field, introduced by Bourdieu (1997), designates a relatively autonomous space, a microcosm endowed with its own laws, some of which are social laws. Each field, or each element of a science–society relationship matrix and TRL, can have partial autonomy (existence of links between the elements of the matrix). One of the problems that must be dealt with concerns knowledge of the

> nature of the external constraints, the form in which they are exercised, credits, orders, contracts, etc., and in what form the resistances which characterize the "autonomy, that is to say what are the mechanisms that the microcosm puts in place to free itself from these external constraints and to be able to recognize only its own internal determinations" (Bourdieu, 1997).

This set of questions, as we will see, is to be addressed, but depending on the nature of the invention and innovation processes:

> In other words, we must get out of the alternative of "pure science", totally freed from all social necessity, and "serve science", subject to

all politico-economic demands. The scientific field is a social world and, as such, it exerts constraints, solicitations, etc., but which are relatively independent of the constraints of the encompassing global social world. In fact, external constraints, of whatever nature, exerted only through the intermediary of the field, are mediated by the logic of the field (Bourdieu 1997).

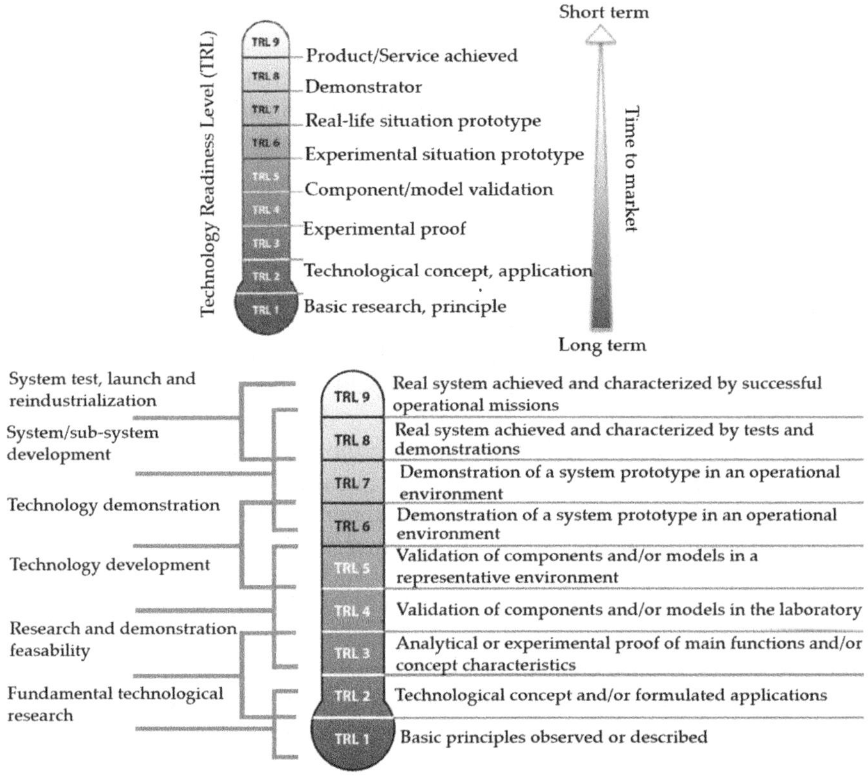

Figure I.9. *TRL (technological maturity level). For a color version of this figure, see www.iste.co.uk/andre/knowledge1.zip*

In addition, Science Europe (2015) suggests, among other things, that when designing innovation policies, the following considerations should be taken into account:

– the challenges faced by contemporary societies will not be solved by incremental developments – rather, they require radically new approaches;

– as challenge-based research has taken the lead over traditional forms of curiosity-based research, this will stimulate the need to build knowledge across disciplines, and thus encourage innovative approaches;

– when radical innovation emerges, it can be a game-changer, with spin-offs beyond those generated by the original challenge;

– is an essential element in applying research-based knowledge to the real world, stimulating profound change.

Decision-makers in charge of innovation should:

– take action across the research ecosystem to support the integration of disciplines and to stimulate genuine interaction between all players;

– make a strong commitment to the cooperative design between policy-makers and researchers from a wide range of scientific backgrounds (from creativity to application through trial and error (including proof-of-concept);

– encourage the implementation of programs focused on innovation, which support all aspects of innovation, and are also fertile for the development of nonlinear and intangible innovation;

– design a new research ecology appropriate to the societal challenges of the 21st century.

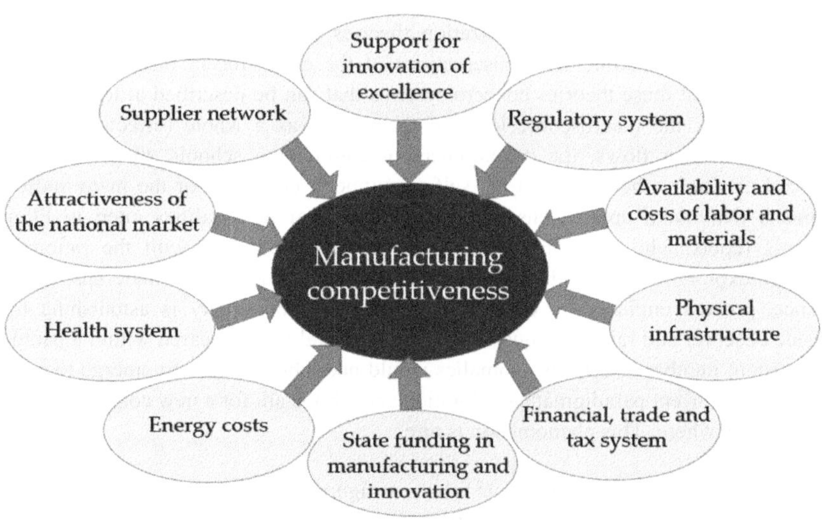

Figure I.10. *Competitiveness development as seen by NNMI. For a color version of this figure, see www.iste.co.uk/andre/knowledge1.zip*

To convince ourselves of the need to act differently to how Brussels is proposing, we need look no further than the United States' NNMI (National Network for Manufacturing Innovation) (Federal Research & Development 2015).

This organization brings together various thematic institutes and is intended to serve as a driver for industrial transfer (values TRL values between 3 and 7). The operating chart is shown in Figure I.10, which is taken from NNMI (2016).

By linking various bodies of knowledge and skills, this structure should strengthen and make more efficient the ways of developing competitiveness (particularly in the field of manufacturing). However, not everything is black, just gray; there are indeed a certain number of companies, rather small ones, which are innovative in this field (Notteau 2016). All of these comments that will be explored in this book will serve as pillars for the analysis of inventions and innovations associated with additive manufacturing and its more or less natural children.

> The vast majority of the Men who run our states, educate our children, uphold the moral order and preserve the treasures of science and wisdom of our ancestors, have been trained in this [traditional] way. They are also the ones who must organize, when necessary, changes in the intellectual path of new generations. They must be encouraged or urged on by the public opinion of the enlightened classes of the whole nation, men and women alike (von Helmholtz 1874).

A contribution from self-organization theories provides a form of description of change that is compatible with observations of the city's recent history. It should be remembered that these theories concern systems that can be described at least two levels of observation: the macroscopic level of the system as a whole (effects of the 1870 war with Alsatian flows, the development of engineering schools, then the return of Alsace-Lorraine to the national fold, self-recruitment, etc.); that of the many individual elements that make up this ensemble (highly original personalities cited in Florence Besset's report, who attempted to create a school of thought with the rational and thorough exploration of paradigms by their successors (mathematics, materials, computer science, process engineering, etc.)). This situation of continuity is astonishing to the outside observer and raises the question of the temporality of research within a paradigm. A persistent number of serious anomalies would undoubtedly have to emerge to call into question the current paradigmatic exploration and, above all, for a new conceptual basis to emerge somewhere. This phenomenon is rare.

In terms of the city, changes in its urban configuration (with an obsessive tourist focus on historic centers), activities (industrialization disappeared for several decades and tertiarization with modest added value), the composition social (increase in local tertiary jobs) and the number of inhabitants (rural exodus and desertification of the periphery, urban decline, loss of intellectual lifeblood more tempted by other sites, etc.) appear as elements of image that cannot sufficiently counterbalance desires for local development, which are not particularly asserted and unsupported. The refuge is the external support with a look at a more glorious past (or considered as such).

In order to understand the global changes, we can take a local look at the flows of individuals, households and companies in a world dedicated to mobility, who make up the city's lifeblood, on a yearly time scale (residential and professional mobility, company start-ups and closures), as well as on a generational scale. With the exception of university staff, for the most part, people (young people in particular) change workplaces fairly frequently, while the urban morphology, described by its physical or social attributes, persists and is transformed over a longer timeframe. We explore the frog syndrome, confident in the glory of its past, perpetuating itself ad infinitum.

Against this ambivalent backdrop, university research and university training systems have evolved in two modes: stable trajectories and a few forks in the road under dynamic conditions (as the world moves on around us). They are continuously transformed by dynamic processes, both internal and external. If a structure emerges and remains persistent enough to be recognizable, it is nonetheless the product of this evolution. Thus, the engineering sciences (including computer science and materials) developed specifically in Lorraine relative to other French sub-regions still constitute a respectable scientific force (even if it is a little out of date for research organizations). However, Figure I.11, which shows the position of Shanghai funding relative to funding from the "Programme d'investissement d'avenir" or PIA funding versus the number of students, shows a clear deficit for Lorraine.

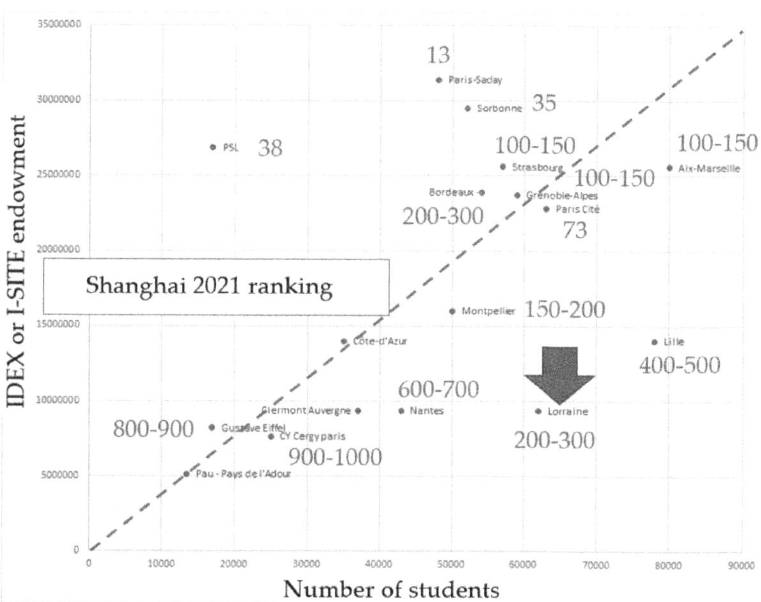

Figure I.11. *PIA funding versus number of students and ranking. For a color version of this figure, see www.iste.co.uk/andre/knowledge1.zip*

Does this mean that teaching takes precedence over research? Finally, in addition to self-organization processes, there is the specific evolutionary nature of a city, which means that the creative capacity of urban centers must also be taken into account, as the flow of innovation and the creation of new wealth are the only factors capable of explaining the continuation of attractive urbanization. How can local science be a real driving force behind this design, which obviously only imperfectly concerns it?

Any local research and innovation system only really makes sense if it is thought through and questioned, at regular intervals, in order to maintain a level of excellence in relation to a certain vision of the future. Nancy's vision of the future is a techno-scientific one (but not the only one). What we're seeing is the daily introduction of increasingly sophisticated mechanisms and procedures for controlling researchers and teacher-researchers, whose operations continue to be business as usual. But the world is changing, with major trends that need to be taken into account, and alternative visions of the future that can/should break with tradition.

The most successful today are open, pluralistic, living, agile and reactive. We speak of "ecosystems", which increasingly implies the integration of randomness, risk-taking, emerging players into decisions, and resourcing actions through science, etc. In fact, by making "objects" ever easier to acquire, by shortening their useful life, by exploiting available low-cost reserves too quickly, consumer society knows it is heading for disaster (even if some believe that technology will remain THE solution in developed countries). So there is a desire for the new, and forms of conscious repentance, lulled into sleep by the vague expression "sustainable development". Is there a desire on the part of the population and its elected representatives to be consumers of these goods, or to participate in some aspects of an economic activity that is still provisional, based on the knowledge of scientists?

Can they rely on a dynamic industrial environment? The answer is clearly no, leading to a local, nonconsumptive academic environment, rather disconnected from the city's economy. Can they rely on the will of the academic community? On a long-term vision? Defining an original societal need and the technical and organizational means to satisfy it does not, of course, fit systematically into the leadership of techno-scientific research. It must retain its scientific character, creativity and depth, while at the same time possessing specific approaches due to its position at the interface with the applicability of scientific concepts derived from research. In fact, the legitimacy of science is built on the dynamic and recursive pooling of scientific knowledge with a view to achieving a goal whose origin lies in the scientific component, at least at national level, or in that of applicators or a request from decision-makers. But then, where are the enlightened decision-makers at local level?

So, to avoid confining the researcher to their discipline, a discipline in which they exercise their freedom (their autonomy) and activities that enable them to go beyond the limits of knowledge, we probably need to find and invent ways of "transgressing", of taking risks which should authorize and support new cultural couplings combining divergence and creativity. These should enable the promotion of synergies, enabling the creation new concepts or artifacts useful to society (notions of competition and performance). This is exactly what is expressed by the (happy?) idea of coupling science and technology. By supporting creativity at the interfaces, by developing creative hybridization, it is therefore a question of going beyond other frontiers of knowledge, but no doubt from new or revisited methods, by resituating research and researchers in the context of city life. This exercise is to be done.

With this aspect, there is a singular element in research, that of rarity, originality, which can induce exchange through the dynamics and attractiveness created. It is interesting to examine why and how exchange relations developed with difficulty internally or externally (in the form of pooling of knowledge, customer–supplier relationship, barter or money), how researchers in their scientific quest have financial and human support (this problem raises the question of the steering of research (or the autonomy of the researcher) and possible cooperation to help some to maintain the course of the "liner" that is for example a large laboratory of research). Indeed, how can the erratic funding experienced by researchers in the support of agencies be smoothed out? So, which forces act to ensure that separate researchers can "hold together" and constitute what academia calls a research unit? Social mediations through which these individual-to-individual relationships are transformed and shaped to avoid anarchy, but to be part of a global compatibility that makes sense, are to be studied and engaged for innovation. What energy is supplied to reach this state of dynamic equilibrium making it possible to fight against the second law of thermodynamics? In fact, does it not clearly appear that the world of research is too self-centered, its promotion and a standardized fetishism of the evaluation that is adapted to it? The relationships between actors then appear in the background, almost in antagonism. Is this misappropriation in relation to the mission (as far as it is defined)? So, what can be the link between researcher, local, national or international economy?

Box I.6. *Well-formed head or good organization?*

Provincial cities like Nancy (the author's hometown, as you would expect) sit quietly on their glorious heritage, but lack daring and leave it to their big sisters (including the capital) to drive the economy. At best, they try to bring in vagrant but evanescent industries to fill job gaps. Some believe that the local university can help redeployment by trusting it and supporting it financially. However, Figure I.12 illustrates the university's limited capacity to ensure sufficient scientific influence (801th place according to QS World University Rankings (2022)), despite a few international nuggets (CNRS, INRIA, INRAE). However, according to the same source, this correct result should be compared with Paris: rankings 44 for Paris-Saclay, 49 for Polytechnique, 72 for Sorbonne Université, or, closer to Nancy (and in the provinces) Strasbourg 421th (see Figure I.11).

One way of thinking about this is to know whether we belong to a research elite scouts and how we can judge the societal impact of tomorrow's creativity. A second is that of the "virtuous followers", wise, learned but compliant. Producers of stabilized knowledge represent the essential backbone of a research unit that needs to rely on common and effectively shared values. A third, which is not opposed to the first, relies on the development of links, both internal and external, disciplinary and/or interdisciplinary. It represents another form of intellectual mobility. Between well-established positions, the merit of scouts and trailblazers can only develop if there is an open window of opportunity, and if there is the human and financial support to explore the uncertain and take real risks.

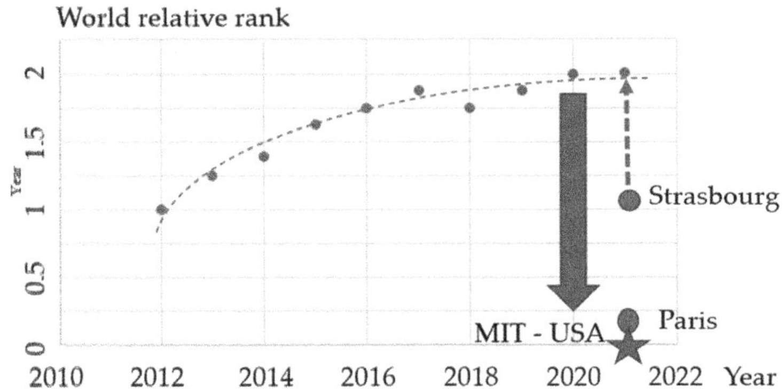

Figure I.12. *Relative evolution of Université de Lorraine's ranking by year, according to QS World University Rankings (2022). For a color version of this figure, see www.iste.co.uk/andre/knowledge1.zip*

So, there is probably no single merit, no single "excellence", but diversity. The richness of a laboratory lies in the diversity and harmony of its skills (and not just its scientific skills), as long as we avoid symbolic balances, non-aggressions and the opaque reality of certain local practices, and calmly and empathetically manage the tensions between these tendencies, which sometimes come under the paradoxical injunction to be "readable", "credible" and "driving": in short, excellent. It is this governance that needs to reinforce a collective imagination to create a space for freedom and joint action, freeing itself, as far as possible, from an "over-riding" administrative technocracy that thinks it is steering research to meet the needs of society (as it imagines them) and continues to apply principles of "meritocracy" adapted to all and perhaps to nothing... But maybe this is the price we have to pay to keep being enterprising, if we know what we are and where we want to go (and how).

The integration of climate change into the agendas will cause certain equipment to become obsolete, which reinforces the need for innovation. This is also what is found in the various EU development plans (Horizon Europe in particular) with funding of around €100 billion over the duration of the plan. For Crevel (2022), Western countries, including the EU, by supporting these major programs in terms of investment, place the developed states at the heart of the battle: "the latter intend to be both planners of energy change, investors and producers of standards. [It is the] return of the master state of the clocks". So, if we use the accelerator and the brake simultaneously, how will the innovation unfold? Especially if, as proposed by Loesekrug-Pietri et al. (2022), the EU relies on inefficient financing procedures? These authors call into question, as in France, "senior civil servants specialists in everything and therefore nothing, calls for projects which are administrative monsters, a sprinkling without priorities or requirement of result". However, in a global economy, the real factors of competitiveness are dynamic, that is to say, they require a permanent process of innovation and improvement that cannot be stereotyped.

Generally speaking, these different "manifestations" of external tensions need to be shared in order to create a common sense of action and a shared collective feeling on the part of the members of the scientific community. Intelligible representations on the part of the entire staff presuppose a minimum of trust between its members, so that they can work together. One possible means of achieving this is a common or partially common scientific project. This material project would not be there to reveal the "mental" state with which the members are associated, but to a means of bringing together energies to do. In order to gain a greater sense of togetherness, it would be a matter of externalizing the movement in order to create links and participate in the evolution of the image, strengthened because all partners have contributed to its creation. In a rational anticipation of uncertainty, this pooling of resources would then have a renewed attractiveness and a strengthened capacity for integration.

But, for the most part, a lack of ideas means that innovation has stalled. This situation is confirmed by Lévy-Provençal (2020):

> At the dawn of the 2020s, reality seems to be taking its time compared to the over-optimistic forecasts of these same technologists. Autonomous cars are slow to arrive, the power of electronic chips is stagnating, artificial intelligence is disappointing, biotechnologies and nanotechnologies are living through their winter [...].

But, at the same time, it could be possible to use this time to initiate locally useful transformations, a possible way of preparing for the future. If R&D is important for growth (bringing innovations to market), "fundamental" scientific research is produced all over the world and is easily accessible via the Internet. Technology transfer is therefore facilitated, provided that policies funding basic research are genuinely committed to fostering the innovations we need locally for long-term growth (FMI 2022).

In this way, it should be possible to demonstrate the value of such an approach to exploiting divergent thinking and to show how we can/should support people in the academic world (or not) who wish to engage in the invention and creating of societal values. Apart from bringing about a useful cultural change, the success of such an operation could lead to a new model of change management. France thinks it is a very big country, and according to the glorious memory of Louis XIV, it is the absence of clear local leadership on the part of local councillors and the presence of the "tertiary depot" constituted by the academic world engaged in goals that do not understand the future of the host city, region or country. This lack of thinking about the future is all the more damaging given that decisions, or rather the absence of decisions, are likely to irreversibly shape the region in the years to come. And when it comes to development phenomena, there is already a visible inertia that can be measured.

Torn between vocation and ambition, between emulation and competition, today's researcher is a somewhat "paradoxical" character, who only partially masters the logic of his actions (thinking about the interactions between scientific knowledge and actions towards/for society to better understand why and how it is possible to adopt a more responsible approach to research, due to very important developments). The quest for a scientific reputation now resembles a sporting competition, a race for "peer-reviewed" publications, overseen by a new culture of research management that more readily evokes the old Soviet centralized planning than the free market of ideas (which we let modestly hatch). Researchers find themselves more than ever a slave to a system they helped to create, and from which they find increasingly difficult to distance themselves.

The scientific world would do well to take a critical look at these new reputational mechanisms, if only to enable researchers to better master the rules, so that they can focus more on their humanist mission than on their "mere" evaluation on criteria that are partly removed from the scientific target and their mission. Evaluation systems have an impact on the way personal and collective reputations are built. The tangle of different tools, standards and practices reveal a mix of esteem and prestige recognition practices (the classic peer review system, the citation system, the economy of greatness based on national and international distinctions, awards and rankings). How does individual performance contribute to the recognition of a research unit's scientific activity (and vice versa)? How can this performance, when it exists, play a valid part in the dynamics of the city? Or, more broadly, society? Clearly, this point of view does not answer the question, even if it does raise questions.

How can we advance/finance proposals for innovative, creative subjects to which research can make a useful contribution? The short-term orientation of what describes the present situation is likely to lead to a "Darwinian" and follow-the-leader lifestyle for researchers, even though they are true professionals, with the survival of those best able to cope opportunistically with the ambient normative universe, but with modest creativity and a target far removed from the future societal needs. The field of 3D/4D/BP printing lends itself particularly well to avoiding this proposition and attempting to think (and act) differently.

> Tell yourself a priori that something is going to change. You set a goal, you make proposals, you sow various seeds everywhere, and you know that time will bring an answer (Moreno 1990).

> The researcher finds himself in the situation of an airplane pilot so absorbed by his instruments that he can no longer turn his head toward the window to see that his engine is on fire (Taleb 2007).

> What exactly is 'thinking'? When, under the effect of sensitive impressions, images arise from the depths of our memory, it is not yet "thinking". When these images are linked in sequences where each term calls for the next, it is still not "thinking". But when a given image recurs in several of these sequences, this image, by the very fact of its repetition, becomes an organizing element, to the extent where it connects series which otherwise would have no connection between them (Einstein 1991).

I.2. References

Accenture (2019). Culture of equality key driver in workplace innovation, report finds [Online]. Available at: https://betakit.com/culture-of-equality-key-driver-in-workplace-innovation-report-finds/.

Adamson, A.W. (1968). Method and apparatus for generating 3 dimensional patterns. US Patent 3609706 A.

Adelman, A.H. and Lewis, J.D. (1968). Method and apparatus for generating 3 dimensional patterns. US Patent 3609707 A.

Aghion, P., Antonin, C., Bunel, S. (2020). *Le pouvoir de la destruction créatrice*. Odile Jacob, Paris.

Alter, N. (ed.) (2002). *Les Logiques de l'innovation*. La Découverte, Paris.

André, J.C. (2015). N'est pas Prométhée ou Mary Shelley qui veut : une petite histoire de la stéréophoto-lithographie... *Actualité chimique*, 397(398), 99–101.

André, J.C. (2016). Désordres dans la troisième dimension ; un cheminement entre formation et interdisciplinarité. *VARIA*, 5, 62–72.

André, J.C. (2017a). *From Additive Manufacturing to 3D/4D Printing 1: From Concepts to Achievements*. ISTE Ltd, London, and John Wiley & Sons, New York.

André, J.C. (2017b). *From Additive Manufacturing to 3D/4D Printing 2: Current Techniques, Improvements and their Limitations*. ISTE Ltd, London, and John Wiley & Sons, New York.

André, J.C. (2017c). *From Additive Manufacturing to 3D/4D Printing 3: Breakthrough Innovations: Programmable Material, 4D Printing and Bio-printing*. ISTE Ltd, London, and John Wiley & Sons, New York.

André, J.C. (2019). *Industry 4.0: Paradoxes and Conflicts*. ISTE Ltd, London, and John Wiley & Sons, New York.

André, J.C. and Corbel, S. (1994). *Stéréo-photolithographie laser*. Polytechnica, Paris.

André, J.C., Le Méhauté, A., De Witte, O. (1984). Dispositif pour réaliser un modèle de pièce industrielle. French Patent no. 84 11 241.

Angeler, D., Sundstrom, S., Allen, C. (2016). Deathcore, creativity, and scientific thinking. *Research Ideas and Outcomes*, 2, e8867-6.

Apolitical (2019). Eight myths about public sector innovation – debunked [Online]. Available at: https://apolitical.co/solution-articles/en/eight-myths-about-public-sector-innovation-debunked.

Balconi, M., Brusoni, S., Oresnigo, L. (2010). In defense of the linear model: An essay. *Research Policy*, 39, 1–13.

Baumol, W. (2002). *The Free Market Innovation Machine: Analyzing the Growth Miracle of Capitalism*. Princeton University Press, Princeton.

Bayrou, F. (2022). Face à la dette, la reconquête nationale de la capacité productive [Online]. Available at: https://www.societal.fr/francois-bayrou-face-la-dette-la-reconquete-nationale-de-la-capacite-productive.

Benedetti, D. (2021). L'émotion de la découverte : une étape de la recherche incitant à la créativité scientifique [Online]. Available at: https://theconversation.com/lemotion-de-la-decouverte-une-etape-de-la-recherche-incitant-a-la-creativite-scientifique-167809.

Bouiss, O. (2020). Qu'est-ce qu'une "crise" ? [Online]. Available at: https://the-conversation.com/quest-ce-quune-crise-136026.

Bourdieu, P. (1997). *Les usages sociaux de la science. Pour une sociologie clinique du champ scientifique*. Quæ, Paris.

Bourdin, S. (2022). Quelle acceptabilité sociale des technologies disruptives ? [Online]. Available at: https://theconversation.com/quelle-acceptabilite-sociale-des-technologies-disruptives-173337.

Bulkeley, H., Coenen, L., Frantzeskaki, N., Hartmann, C., Kronsell, A., Mai, L., Marvin, S., McCormick, S., Steenbergen, F., Voytenko Palgan, Y. (2016). Urban living labs: Governing urban sustainability transitions. *Current Opinion in Environmental Sustainability*, 22, 13–17.

Case, A. and Deaton, A. (2021). *Morts de désespoir – L'avenir du capitalisme*. PUF, Paris.

Christensen, U.J. (2021). Greater competitiveness tops the goals for 2022; how thinking can help [Online]. Available at: https://www.forbes.com/sites/ulrikjuulchristensen/2021/12/03/greater-competitiveness-tops-the-goals-for-2022-how-thinking-can-help/?sh=142cb2605424.

Cinquin, P., Kenzari, S., Dufour, A., Lefebvre, S., Demoly, T., André, J.C. (2021). Recherches et fabrications additives pour la santé – Mutualisations franco-africaines ? [Online]. Available at: https://www.techniques-ingenieur.fr/actualite/livre-blanc/recherches-et-fabrications-additives-pour-la-sante-mutualisations-franco-africaines-94526/.

Crevel, P. (2022). État, le retour après une fausse sortie [Online]. Available at: https://www.lesechos.fr/idees-debats/cercle/opinion-etat-le-retour-apres-une-fausse-sortie-1381645.

Demoly, F. and André, J.C. (2022). *4D Printing 1: Between Disruptive Research and Industrial Applications*. ISTE Ltd, London, and John Wiley & Sons, New York.

Deserti, A., Real, M., Schmittinger, F. (2022). *Co-creation for Responsible Research and Innovation*. Springer, Cham.

Devillers, L. (2022). Pour un Ministère de l'Innovation numérique positive et responsable [Online]. Available at: https://www.acteurspublics.fr/articles/laurence-devillers-pour-un-ministere-de-linnovation-numerique-positive-et-responsable.

DGA (2009). Quelques explications sur l'échelle des TRL (Technology Readiness Level) [Online]. Available at: https://www.entreprises.gouv.fr/files/files/directions_services/politique-et-enjeux/innovation/tc2015/technologies-cles-2015-annexes.pdf.

Dortier, J.F. (2015). Innover… contre les idées reçues. *Les grands dossiers des sciences humaines*, 38, 1.

Dubois, M.J.F. and Brault, N. (2021). *Manuel d'épistémologie pour l'ingénieur(e)*. Éditions Matériologiques, Paris.

Dufourmentelle, A. (2011). *Éloge du risque*. Payot, Paris.

Einstein, A. (1991). *Œuvres choisies*. Le Seuil/CNRS, Paris.

ERC (2018). ERC Work Program 2019 [Online]. Available at: http://ec.europa.eu/research/participants/data/ref/h2020/wp/2018-2020/erc/h2020-wp19-erc_en.pdf.

EU (2018). Science, research and innovation performance of the EU, 2018 – Strengthening the foundations for Europe's future [Online]. Available at: https://op.europa.eu/en/publication-detail/-/publication/16907d0f-1d05-11e8-ac73-01aa75ed71a1/language-en.

Fiasco (2022). Une revue dédiée aux projets inachevés, avortés, oubliés, ratés… [Online]. Available at: https://www.kisskissbankbank.com/fr/projects/fiasco-le-monorail.

FMI (2021). De l'importance des sciences pures pour la croissance économique [Online]. Available at: https://www.imf.org/fr/News/Articles/2021/10/06/blog-ch3-weo-why-basic-science-matters-for-economic-growth.

FMI (2022). Pourquoi la recherche et le développement sont un élément crucial de la croissance économique [Online]. Available at: https://www.weforum.org/agenda/2021/10/basic-science-economic-growth-collaboration.

Freeman, C. (1988). À quoi tiennent la réussite ou l'échec des innovations dans l'industrie ? [Online]. Available at: http://documents.irevues.inist.fr/bitstream/handle/2042/31862/C&T_1988_18_30.pdf?sequence=1.

Friedman, G. (1952). Les conséquences sociales du progrès technique. *UNESCO – Bulletin international des sciences sociales*, 4, 251–272.

Gaudin, B. (2008). Innovation: The history of a category. Working paper no. 1 [Online]. Available at: http://espace.inrs.ca/id/eprint/10023/1/Godin_2008.pdf.

Godin, B. (2006). The linear model of innovation: The historical construction of an analytical framework. *Science, Technology, and Human Values*, 31, 639–667.

Grisoni, P. (2022). *Fiasco, l'art de rebondir après l'échec*. Hugo Doc, Paris.

Hammond, M.E.H., Stehlik, J., Drakos, S.G., Kfoury, A.G. (2021). Bias in medicine. *JACC: Basic to Translational Science*, 6, 78–85.

Harari, Y.N. (2018). *Sapiens: A Brief History of Humankind*. HarperCollins, New York.

Heidegger, M. (1968). *What is Called Thinking?*, translated by J. Glenn Gray. Harper & Row, New York.

von Helmholtz, H. (1874). Sur l'ambition de vulgariser la science. *Philosophia Scientiae*, 9(1), 59–71.

Herbet, M.E., Léonard, J., Santangelo, M.G., Albaret, L. (2021). Dissimuler ou disséminer ? Une étude sur le sort réservé aux résultats négatifs [Online]. Available at: https://zenodo.org/record/5541589?mc_cid=feb168606a&mc_eid=eb160fe7cc#.YWRm1NWxXtS.

Hull, C. (1984). Method for production of three-dimensional objects by stereolithography. US Patent 5,762,856.

IREF-Europe (2022). Réindustrialiser la France… un horizon très lointain [Online]. Available at: https://www.contrepoints.org/2022/02/14/421588-reindustrialiser-la-france-un-horizon-tres-lointain?utm_source=rss&utm_medium=rss&utm_campaign=reindustrialiser-la-france-un-horizon-tres-lointain.

Jakobsen, S.E., Fløysand, A., Overton, J. (2019). Expanding the field of Responsible Research and Innovation (RRI) – From responsible research to responsible innovation. *European Planning Studies*, 27, 2329–2343.

Kasbarian, G. (ed.) (2019). 5 chantiers pour simplifier et accélérer les installations industrielles – Pacte productif 2025 [Online]. Available at: https://www.gouvernement.fr/sites/default/files/document/document/2019/09/rapport_kasbarian.pdf.

Koestler, A. (1976). *The Ghost in the Machine*. Random House, New York.

Koyré, A. (1988). *Du monde clos à l'univers infini*. Gallimard, Paris.

Krichen, K. and Chabaud, D. (2019). Une analyse comparée et critique des modèles des spin-offs universitaires. *Marché et organisations*, 34, 61–86.

Landes, D.S. (2006). Why Europe and the West? Why not China? *The Journal of Economic Perspectives*, 20, 3–22.

Learning Technologies (2022). Stefaan van Hooydonk [Online]. Available at: https://www.learningtechnologies.co.uk/speakers/stefaan-van-hooydonk.

Leleu-Merviel, S. and Boulekbache, H. (2014). *Recherches en design – Processus de conception, écriture et représentations*. ISTE Éditions, London.

Lévy-Provençal, M. (2020). Qui seront les prospectivistes de demain ? [Online]. Available at: https://www.lesechos.fr/idees-debats/sciences-prospective/qui-seront-les-prospectivistes-de-demain-1166587.

Loesekrug-Pietri, A., Guillou, M., Henri, D., Vandewalle, T., Verre, L. (2022). Innovation : Monsieur le président, faites tomber ces murs ! [Online]. Available at: https://www.lesechos.fr/idees-debats/cercle/opinion-innovation-monsieur-le-president-faites-tomber-ces-murs-1386396.

Maddyness (2022). 3 pistes pour doper l'innovation technologique française [Online]. Available at: https://www.maddyness.com/2022/03/21/3-pistes-doper-innovation-technologique-francaise/.

Mansfield, E. (1998). Academic research and industrial innovation: An update of empirical findings. *Research Policy*, 26, 773–776.

Mapping-Consulting (2012). Liens entre les technologies clés génériques, les filières industrielles, les échelles TRL, nos défis sociétaux. Business conditions.

Markham, S.K. (2002). Moving technologies from lab to market. *Research-Technology Management*, 45, 31–42.

Maslow, A. (1966). *The Psychology of Science*. Harper & Row, New York.

McGinniss, V.D. and Schwerzel, R.E. (1984). Photo-polymerizable composition containing a photo-sensitizer donor and photo-initiator acceptor. US Patent 4571377 A.

MEF (2022). Communiqué de presse – La France et l'Allemagne s'engagent à investir 1 milliard d'euros chacune dans le futur fonds ETCI (European Tech Champions Initiative) [Online]. Available at: https://presse.economie.gouv.fr/08-02-2022-communique-de-presse-la-france-et-lallemagne-sengagent-a-investir-1-milliard-deuros-chacune-dans-le-futur-fonds-etci-2/.

MESRI (2022). Place de la R&D française dans l'UE [Online]. Available at: https://www.enseignementsup-recherche.gouv.fr/fr/place-de-la-rd-francaise-dans-l-ue-83080#.YeFElLTS5Ok.twitter.

Mokyr, J. (2017). The past and the future of innovation: Some lessons from economic history [Online]. Available at: https://www.nber.org/system/files/chapters/c14016/revisions/c14016.rev0.pdf.

Montenay, Y. (2022). On manque de bras et de cerveaux, où sont-ils ? [Online]. Available at: https://www.yvesmontenay.fr/2022/01/19/on-manque-de-bras-et-de-cerveaux-ou-sont-ils/.

Moreno, R. (1990). *Théorie du bordel ambiant – Souvenirs de l'irréversible*. Belfond, Paris.

Naudon, F. (2013). Comment le profane joue en faveur du décloisonnement. *Hermès, La revue – Interdisciplinarité : entre disciplines et indiscipline*, 67, 62–67.

Normand, F. (2022). Le Canada veut (re)devenir un leader mondial dans les sciences de la vie [Online]. Available at: https://www.lesaffaires.com/blogues/francois-normand/le-canada-veut-redevenir-un-leader-mondial-dans-les-sciences-de-la-vie/630860.

Nowotny, H. (2006). The quest for innovation and cultures of technology. In *Cultures of Technology and the Quest for Innovation*, Nowotny, H. (ed.). Berghan Books, New York.

Nowotny, H. (2008). *Insatiable Curiosity: Innovation in a Fragile Future*. MIT Press, Cambridge.

Oakey, R. (2015). Schumpeterian economics: Some observations on the relevance of his theoretical contributions to the management of industrial research and development. *R & D Management*, 45, 386–396.

OECD (2021). Knowledge co-creation in the 21st century – A cross-country experience-based policy report [Online]. Available at: https://www.oecd-ilibrary.org/docserver/c067606f-en.pdf?expires=1642087314&id=id&accname=guest&checksum=BE37307165D496F421A0B78CF12BA91D.

Popper, K. (1973). *Logique de la découverte scientifique*. Payot, Paris.

Porter, M. (1980). *Competitive Strategy: Techniques for Analyzing Industries and Competitors*. Free Press, New York.

Porter, M. (1985). *Competitive Advantage: Creating and Sustaining Superior Performance*. Free Press, New York.

QS World University Rankings (2022). Université de Lorraine, Nancy [Online]. Available at: https://www.topuniversities.com/universities/universite-de-lorraine.

Ridley, M. (2020). *How Innovation Works: And Why It Flourishes in Freedom*. Harper, New York.

Röpke, W. (2022). Faut-il créer un ministère de l'innovation ? [Online]. Available at: https://philippesilberzahn.com/2022/02/14/faut-il-creer-un-ministere-de-l-innovation/.

Rosa, H. (2010). *Accélération*. La Découverte, Paris.

Saint-Étienne, C. (2015). *La France 3.0 – Agir, espérer, réinventer*. Odile Jacob, Paris.

Schmookler, J. (1962). Economic sources of inventive activity. *The Journal of Economic History*, 22(1), 1–20.

Schumpeter, J.A. (1934). *The Theory of Economic Development: An Inquiry into Profits, Capital, Credit, Interest, and the Business Cycle*. Harvard University Press, Cambridge.

Schumpeter, J.A. (1947). The creative response in economic history. *Journal of Economic History*, 7, 149–159.

Silberzahn, P. (2022). Faut-il créer un ministère de l'innovation ? [Online]. Available at: https://philippesilberzahn.com/2022/02/14/faut-il-creer-un-ministere-de-l-innovation/.

Simon, H.A. (1976). From substantive to procedural rationality. In *Methods and Appraisal in Economics*, Latsis, S.J. (ed.). Cambridge University Press, Cambridge.

Solow, R.M. (1957). Technical change and the aggregate production function. *Review of Economics and Statistics*, 39(3), 312–320.

Spencer, H. (2017). *First Principles*. John Childs and Son, London [Online]. Available at: https://www.gutenberg.org/cache/epub/55046/pg55046-images.html.

Stiegler, B. (2021). *Nietzsche et la vie – Une nouvelle histoire de la philosophie*. Gallimard, Paris.

Taleb, N.N. (2007). *Le cygne noir : la puissance de l'imprévisible*. Les Belles Lettres, Paris.

Taylor, F.W. (2006). Report of a lecture by and questions put to Mr. F.W. Taylor: A transcript. *Journal of Management History*, 14, 214–236.

The Innovation Research Interchange (2022). 2022 R&D trends forecast. *Research-Technology Management*, 65, 18–28.

Tibbits, S. (2013). The emergence of 4D printing [Online]. Available at: https://www.ted.com/talks/skylar_tibbits_the_emergence_of_4d_printing.

Tibbits, S. (2014). 4D printing: Multi-material shape change. *Architectural Design*, 84, 116–121.

Tsui, A.S. (2007). From homogenization to pluralism: International management research in the academy and beyond. *Academy of Management Journal*, 50, 1353–1364.

Tsui, A.S. (2021). Responsible research and responsible leadership studies. *Academy of Management Discoveries*, 7, 166–170.

Valéry, P. (1957). *Œuvres*. Gallimard, Paris.

Verdier-Molinié, A. (2022). *Le vrai état de la France*. L'Observatoire, Paris.

Vicard, V. and Bensidoun, I. (2022). Réindustrialiser, mais pour quoi faire ? [Online]. Available at: https://theconversation.com/reindustrialiser-mais-pour-quoi-faire-176810#:~:text=Pour%20autant%2C%20une%20politique%20de,trimestre%202021)%20et%20son%20d%C3%A9clin.

Wilson, G.A. (2022). 3 ways for businesses to fuel innovation and drive performance [Online]. Available at: https://theconversation.com/3-ways-for-businesses-to-fuel-innovation-and-drive-performance-174276.

Wittrock, C., Forsberg, E.M., Pols, A., McNaghten, P., Ludwig, D. (2021). *Implementing Responsible Research and Innovation: Organizational and National Conditions*. Springer International Publishing, Cham.

Zhou, C. and Wang, R. (2020). From invention to innovation: The role of knowledge-intensive business services in technology commercialization. *Technology Analysis & Strategic Management*, 32, 1436–1448.

1

Invention: Creativity to Proof of Concept

1.1. Introduction: concepts and the innovation "valleys of death"

Many things have been written to try to explain how and why a scientific, technical, cultural, economic, etc., approach to creativity is useful and essential. To understand the interdependent processes involved, it is probably helpful to have a good grasp of how the most creative people work and think, and how their creative productivity can be fostered all the way to innovation. In this context, it is important to also understand how organizations support, integrate and exploit creative ideas. If it is credible to know the relationship between the market and creativity and how to adapt state and corporate policies to amplify this process, this part (Chapters 1 and 2 of Volume 1 and Chapter 1 of Volume 2) highlights a number of factors which, one after the other and together, have a negative impact on industrial development. They are called "valleys of death" and there are 12 in number. However, the first part (Chapters 1 and 2 of Volume 1 and Chapter 1 of Volume 2) also evokes the winning paths of crossing these trials, depending on the singular situations mentioned, if not in depth. In light of recent economic crises (pandemic linked to Covid-19; "war" between Russia and Ukraine), not very optimistic heavy trends (global warming, depletion of reserves, etc.), creativity, invention and innovation are the basis of a new ideology of industrialization and/or reindustrialization under the constraint of a new "fuel" of a new economy.

An "idea" is often defined as a mental representation of an original solution to a known or new problem (Maunoury 1968) or an intention for action (Parmentier et al. 2017). "Creativity escapes overly rigid structures, plays on inertia and opposes conservatism. It does not consist of perfecting knowledge and techniques, but in making them obsolete by creating new ones. Creativity is sensitive, in the sense that it involves the senses and favors immediate perception" (Aubin 2013). An unusual question may be formulated without a spontaneous answer, but according to

For a color version of all the figures in this chapter, see www.iste.co.uk/andre/knowledge1.zip.

Bonasso (2007), it can lead us to a new way of thinking and being in our work environment: “And it is in the fact of asking this question that we further introduce the new ingredient and begin to change how we are. When we inquire we are present in a different way than our usual awareness. This brings us to the first stage of the process”. But consequential invention is a source of the “law of unintended consequences” (with a happy performative). The idea assumes that we are able to exploit interpretive tools already present to make sense of new circumstances. So, we can selectively choose interpretive frameworks to solve problems and make sense of unusual things, which sometimes leads to the right, actually promising, idea.

At this stage, the idea is not materialized by anything, but it can carry rupture, incrementation, hopes, or wasted time, etc. One of the conditions of creativity is simultaneously deepening, but also widening.

> Deepening that would not involve broadening would perhaps lead to scholarship, never to culture. That’s the difference between a narrow mind and a broad mind. However, if we stick to the pedagogical aspect, this enlargement has a name, it is interdisciplinarity, in other words the reciprocal contribution, in the same field, of various disciplines. The more we advance in the frequentation of a science, the more we are led to call on other sciences until the moment when the combination of these different sciences is so complex that it gives birth to a new science (anonymous reference from Cramarégeas (2021)).

Then, the selection of ideas is an evaluation problem to be carried out as a priority to identify the most promising ones. What interests us here is to consider it as a strategic resource (Harvey et al. 2013), or as a trigger for the innovation process (Ininou and Loilier 2021). It can be considered as the result of social relations (Csikszentmihalyi and Csikszentmihalyi 2006; Vandenbosch et al. 2006). So, the idea is a combination of current knowledge (Van de Ven 1986).

Larousserie (2022), in an interview with Antoine Petit, head of the CNRS, and speaking about invention, writes:

> The race to be the first to invent, to “be the first”, stimulates a short-termism in research that is contrary to its very nature. Any discovery is more often than not the fruit of a long community process, which eventually springs to life under the impetus of one or another.

This comment deserves analysis, because we can have a new idea because we are aware of a scientific and/or technological barrier that can effectively be resolved in the short term, but access to this knowledge can be the result of a scientific culture (not necessarily disciplinary) that shows how scientists interact with the world that supports them, with a reality that takes them out of the disciplinary silos.

The Chairman commented on this aspect, saying: “Every researcher wants to be the first. That’s the principle of research. Research is the discovery, the invention of things that others haven’t done. So, by definition, it’s a form of competition. You have to accept that”.

Admittedly, this is slightly simplistic, as research has a much broader remit. But if they are paid by the nation, researchers can naturally embark on this path, as long as we do not neglect the essential elements that will enable them to reach this level of elitism, by providing opportunities for in-depth study upstream, via support for this elitism downstream, particularly in the technical field. As someone who has worked with colleagues to open up technology to society, I can testify that there is undoubtedly still some thinking to be done before we reach this desirable operational state of valorization of research!

But a scientific fact remains a proposition whose relevance is not questioned. Once accepted, it is only used by other researchers as part of other reticular processes of scientific fact creation (accumulation, extension). This fact must respect the operating principles of science and, at the same time, be included in a network of other accepted facts to gain legitimacy. Thus, for Brisset (2014), the objectification of the fact is the result of the slow construction of a network of allies. What may be true for science under construction is less true for invention, which needs to serve the competitiveness of companies. Callon and Ferrary (2006), moreover, write, as a sign of defining the paths between these two considerations:

> Economic agents are calculators because they are equipped, fitted out, endowed with a host of prostheses. They choose rationally, but follow algorithms defined by others, a bit like when a chess player falls under the calculating sway of his opponent.

Against this backdrop, we need to succeed in creating elements of economic desirability through various means of creation and mobilize financing for innovation *startups*, which is essential for developing tomorrow’s economy (France Angels 2022). According to this source, investment in technology, measured in euros per capita between 2017 and 2021, amounted to €1,567 in Sweden, €1,076 in the UK, €878 in Ireland, €558 in Denmark, €391 in Germany and €383 in France. If there is room for improvement on the financing side, this book focuses on the other side, that of the link between the flow of creativity and the provision of a shared (and, if possible, profitable) technological offering.

Silverzahn (2022) reminds us of some of the principles, laid down by Smith (2000), of the relationship between production and society, presented as follows:

– prudence (calculation, self-interest, ensuring continuity);

– temperance (not taking excessive advantage);

– justice (mutually beneficial relationship);

– hope (gain, confidence, reputation);

– job satisfaction (interactions, exchanges);

– courage in our work;

– faith in activity (understood as something linked to the transcendent, that is, not reducible to a calculation of our interest, pride in our work, social recognition).

This reminder of the need for socially responsible research (which will be the subject of Chapter 1 of Volume 2) is presented here, because this link between creativity and industrial development cannot be determined by technical and financial criteria alone, and it is therefore important for the research community to be able to draw on the foundations of cultural evolution to better associate science and society (André 2013). If we fail to take into account the social relationships associated with application-oriented scientific discoveries, as Brisset (2014) reminds us, we run the risk of falling back into the ways of the Whigs (the dominant liberal party in the UK between the 17th and 19th centuries), "that is, making the history of victors and dominant social groups by considering scientific devices only in their orthodox social uses".

Thus, in the introduction to this book, the transition between an idea and its formalization was recalled, as it enables a transition to innovation via invention, which must already illustrate real applicative potential via proof of concept (POC). This chapter therefore takes into consideration the transition from a purely intellectual process (even if the know-how can be used to frame the successful idea that leads to invention) to an object in the form of a POC. Remember that:

> Imagination consists of imagining absent objects and combining images of these objects with one another. It virtually brings new possibilities into existence. In this sense, it frees us from frameworks and leads us to see the world from a different angle, breaking away from the dominant logic. It allows us to experiment with a wide range of possibilities, leading to new ideas (Parmentier 2020).

But, on the one hand, as Cowen (2011) reminds us:

> [...] we are failing to understand why we are failing. All of these problems have a single little-noticed root cause: We have been living off low-hanging fruit for at least three hundred years [...]. Yet during the last forty years, that low-hanging fruit started disappearing, and we started pretending it was still there. We have failed to recognize that

> we are at a technological plateau and the trees are more bare than we would like to think.

On the other hand, the constrained, if not accelerated, rhythms of life should indicate that we are plunged into instability. The situation forbids any reassuring belief; the landmarks that old societies had built for themselves are becoming obsolete. So, for innovation, we need to invest, as quickly as possible, in a field that is fairly untouched in terms of good methods, including that of complexity without, at this stage, taking the social field fully into consideration. But, along with invention comes confidence!

Complexity is a situation with a variety of different elements, forms and natures in dynamic interaction. Innovation moves the complexity (without necessarily exploring or exploiting it fully) of a situation by modifying a relationship between interacting elements. It can be disruptive or incremental in origin, with characteristics of interaction between multiple players, interdisciplinarity, uncertainty, unpredictability, co-evolution of the project and its context, and so on. Some believe it requires an evolving strategy to emerge. It is this vision that we may wish to question based on our innovations:

– more traditional but whose societal future – from multiple technological, supply criticality, economic and social attractiveness points of view – explores the complexity (recursivity, interdependencies, nonlinearities, etc.);

– breakthroughs involving the difficult convergence of separate disciplines;

– financial and human support logics that are increasingly based on current fashion and the short term.

The National Center for Science and Engineering Statistics (NCSES) (2021), member of the National Science Foundation, has sought to better understand the causes and patterns of innovation in companies. The causes of failure of products or business processes developed and brought to market are defined as follows:

– there was little or no demand;

– the product or business process did not perform as expected;

– the adjacent business inputs needed for customers to use it successfully were not sufficiently available;

– regulatory restrictions on its use have limited its economic value;

– a product or business process has been developed, brought to market and generated positive interest. However, the innovator has not been able to ensure a solid positioning of the innovation in the market, for example, through the protection of intellectual property.

On this basis, following the example of asbestos – the former magic mineral – (whose carcinogenic effects had been deliberately set aside) with its reasoned use, we moved on (with good reason) to its rejection because the unacceptable had emerged, imposing lateralizations (artificial fibers), then applying the precautionary principle. On the other hand, there are no anti-digital verses in the media. So it is not all straightforward in the complex space between "simple" and social agreement, and between invention linked to imagination and innovation. There are constraints, not all of them scientific or technical in origin.

Techno-scientific reality is a complex phenomenon, combining strong demands for business competitiveness with a continuous flow of intentions, actions, events, experiences, determinations and hazards. This sense of complexity has grown, despite the omnipresence of scientific disciplines in the field of innovation, in the face of changing environments outside the world of research. However, creative action is struggling to find effective support in a world that is increasingly constrained or standardized, subject to institutional management methods that are more or less unsuitable, a world confronted with acute problems of governance, saturated with a priori management and evaluation tools.

Thinking about complexity implies, in a universe of increasingly nonlinear causalities, that we probably need to adopt a less classical vision, and therefore one that is closer to humility in human actions, in order to admit that the action of disruptive innovation, once caught up in unstabilized interdisciplinary and organizational dynamics, may in part escape the initial intention. Against this shifting backdrop, the desire to control, rationalize, standardize behavior, reduce risk-taking applied to uncertain systems and increasingly complex systems, reinforces the risk of failure and, potentially, their difficult governance in an unsteady world by obscuring disciplinary hierarchies, tensions, questioning, understanding of the oppositions between process and content and the qualities of interdisciplinary interactions that are so essential to their success.

For Wallas (1926), creativity can be represented by an uncontrollable black box, which originates in sudden "illuminations" appearing in the minds of certain individuals after a period of "incubation" and situational setting, taking into consideration the knowledge of the real and the possible. It arouses, brings into existence and makes desirable other ways of paying attention, making sense and/or giving new importance to things. But, as long as we remain at the stage of the idea or concept stage, an invention is only proven if it exists(!), and it is generally not feasible to take out a patent.

The purpose of a patent is to protect a technical innovation, that is, "a product or process that provides a new technical solution to a given technical problem", according to the French National Institute for Industrial Protection (INPI), quoted by JDN (2021). It is also a means of assessing inventive activity in industrialized countries (Aghion et al. 2020). However, this patent is only a basis for financial negociation within a legal framework (see the claims in the patent drafting), but what may be of interest to the practitioner is the POC that gave substance to the idea.

NOTE.– Einstein and patents:

A patent office collects many technological innovations that need to be examined with great intellectual rigor and a sharp critical eye. Heitz (2002) quotes Albert Einstein: "A profession with practical ends is a blessing for a man like me; an academic career obliges young researchers to have a scientific output, and only strong characters can resist the temptation of superficial research". This remark is important in that it equates scientific research and invention.

Figure 1.1 (Mandard 2018) shows that patents are elements of evaluation. They are likely to identify sources of invention within an organization. "Since they are granted by patent offices that follow standardized procedures, they are indeed a valued indicator".

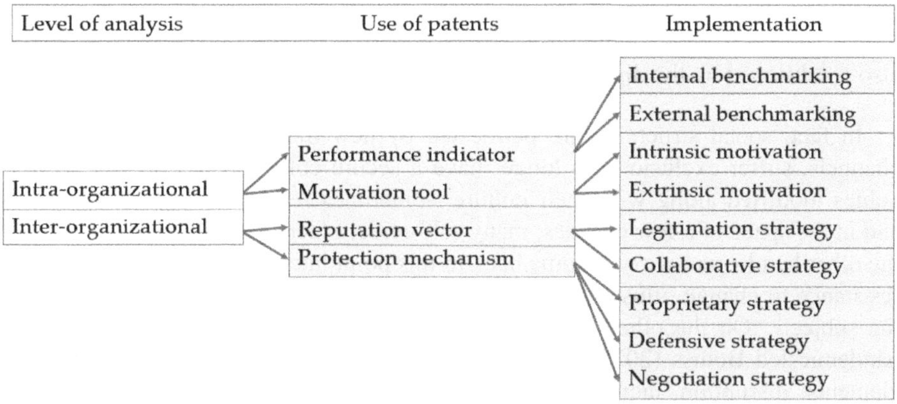

Figure 1.1. *Patents as a factor in reputation and performance*

This chapter is therefore divided into two main sections: one focusing on creativity (if possible, that which leads to technological disruptions) and the other focusing on the epistemological and, above all, practical realization of POCs.

1.1.1. *A look back at creativity*

> Every kind of Evolution is from a state of indefinite incoherent homogeneity to a state of definite coherent heterogeneity (Spencer 2017).

> There are still harmless self-observers who believe that there are "immediate certainties"; for instance "I think" or [...] "I will" (Nietzsche 2013).

> We are dealing with ideology whenever a signifying production is considered in relation to the basic mechanisms of social functioning as constraints on the generation of meaning (Véron 1979).

"We are in the midst of a long and arduous process leading to the emergence, in one form or another, of a global society whose probable structure it is not yet possible to imagine" (King 1999). Recent trends, which the decline of a certain balance of terror during the Cold War has allowed, are reflected in the multi-faceted phenomena of a Europe-wide association of the globalization of trade on the one hand, with fragmentations in production, but also with environmental concerns, the re-emergence of disjunctive forces, disenchantment, irrationality, xenophobia and other forms of fundamentalism, inward-looking attitudes, competition, etc., on the other hand. In short, the future traced by the liberal system is no longer as certain as it was yesterday, which calls for a look at possible strategies that condition and are also conditioned by the digital world, which can no longer be ignored.

In fact, social structures are permeated by new immediate digital information channels, suffer exclusion, no longer have a secure collective vision and see their values modified along with their culture, sometimes reaching a sterile uniformity, and in the space of a few decades, many self-evident facts have become obsolete. On the other hand – and we will come back to this point later – we can observe forms of resistance to change, illustrated, for example, by Blumenberg's (1993) statement on the subject: "On the other hand, the global crisis brought about by the Covid-19 pandemic led Bouiss (2020) to comment on the current crisis, introducing several elements: disruption, increased disorder and uncertainty, blockage and a return to equilibrium". During the first crisis, the author was able to participate in small-scale creativity made possible by a certain absence of hierarchies (André 2020a, 2020b). "The inflationary outlook, the non-redundancy of certain sectors of the economy, the risk of new waves of Covid-19, the geopolitical risks arising from the movement of geopolitical tectonic squares, the neo-populist danger, etc., were in fact certainly anticipated by decision-makers in the public and private sectors, who suddenly had both more time and more money to endorse their strategies" (GEAB 2022). As there were reserves, the pandemic did not lead to major disruptions in the economic recovery as such.

We are three decades away from 2050, and in all likelihood the world will be a very different place than we can imagine by the mid-century. A number of factors need to be taken into account when estimating future needs based on new scientific paradigms, the field of economic possibilities, social aspects with ideological, political and environmental desirabilities and rejections, and so on.

In his inaugural address, President Roosevelt (quoted by Pigasse (2014)) said, “Their efforts bore the stamp of a tradition. They came to exhortations, pleading with a tear in their eye for a return to confidence. They knew only the rules of a selfish generation. They had no vision, and without vision, the people die”. This observation of sharing induces a certain robustness to projection on the future. This robustness is essential to research, because time is an important element to take into account (provided we have a certain vision of the future to be built). The complexity is at work, with paradoxical injunctions and uncertainties, and within a constrained framework, it is important not to make too many mistakes in your choices (Bauman 2006).

However, we need to take into account the different timeframes involved: technological and organizational innovations that are currently increasingly emerging, the pace of life as we accept the new and cultural changes. This asynchronous aspect addressed by Revault d’Allonnes (2012) translates into losses of frames of reference, which are likely to lead to misunderstandings, and even rejections and crises, especially in a possible logic of degrowth (which is rarely taken into account (Briens 2015; Godefridi 2021)). Perhaps then, incremental or disruptive innovators will, in an interdisciplinary approach, need the now essential insights of specialists in the human and social sciences.

What will we do if they have not been considered in the LPPR (*Loi de programmation pluriannuelle de la recherche* (French multi-year research program) simplified to LPR (without the multi-year))? “Our civilization is struggling to extricate itself from the fascination of the past. It only dreams of the future, and when it draws up projects that are no longer mere dreams, it draws them on a canvas where it is still the past that is projected. It is stubbornly retrospective. It needs to become ‘prospective’” (Goux-Baudiment 2002). Prospective memory, however, is that of programmed intentions (it enables us to remember to carry out a foreseeable action, which we can integrate into a plan or program). It is this dark side of technological future management of the technological future, which does not allow a more liberated creativity, which would need to be recognized as such, and consequently financed, and that does not fit into programming!

However, today, knowledge and its unpredictability are recognized by companies as drivers of innovation with the imperative of integrating them into their know-how and their rapid, efficient use. System integration is made necessary by complex technological developments and a greater number of inputs than in the past (Malecki 2010). In principle, knowledge as an input and output can provide a framework for economic activities by distinguishing between applications of classical principles (diminishing returns) in classical manufacturing (Arthur 1996) and those of the "world of increasing returns" based on disruptive knowledge, which generates instability and cumulative causality: thanks to creativity and its fulfillment in the form of invention, we are entering a more unstable world, because we do not know how to program invention (but how to enable it), with the risk of lock-in and possible predominance of an inferior product, or even unanticipated profits for the winner.

There are different ways of characterizing the world facing knowledge-based companies: a new economy (Floricel and Miller 2003), an open economy (Chesbrough 2003; Chesbrough and Crowther 2006) and a meta-national economy (Doz et al. 2001). All these terms refer to the imperative for companies to cope with hyper-competition (D'Aveni 1994) and avoid the "innovator's dilemma" of failing to anticipate disruptive technologies (Christensen 1997). Rias (2017) points out that a high level of control and evaluation inhibits the ability to explore. For companies engaged in this process, there is a need to strike a balance (Gibson and Birkinshaw 2004). These authors write: "Managers must also have long-term orientations through creativity and risk-takings to support exploratory activities" in order to encourage exploration with a high degree of freedom. If this statement has anything to do with the creativity associated with innovation (Amabile et al. 1996; Ford 1996), the recent period linked to the Covid-19 pandemic, reflected in a number of speeches advocating for the liberalization of research, creativity and innovation, has not translated into exceptional results (André 2020a). In fact, we are increasingly witnessing the creation of a *start-up* which, once it has found its market, is bought out by a larger company. It is a way of innovating through "intermediary companies", where the big fish eat the smaller ones.

1.1.2. *Creativity, disruption and crisis*

> Where you are when you are doing what you are doing matters (Gero 1998).
>
> Contradiction, as Plato knew, is the sole instrument of developing thought. But there is a legitimate and an illegitimate use of contradiction. The illegitimate use consists of combining incompatible assertions, as if they were compatible (Weil 1958).
>
> Sapere aude (dare to know) (Despret 2019).

Valley of death no. 1: having a "good idea"

The imaginary is opposed to the real. It is a fragile trace conferred by the absence of a presentable object ("the enemy of reason, which delights in controlling and dominating it" (Pascal 2000)). Although it is not easy to determine whether creativity is based on innate predispositions and/or whether a learning strategy can be taught or guided to promote it, what Fei and Lin (2013) do show is the validation of the cognitive theory thesis that the way an individual processes information has an impact on their creativity. These authors extended the boundaries of the theory to include the type of search with visual and schematic stimuli, showing that the importance of depth of knowledge search and curiosity was associated with highly creative students rather than less creative ones.

Creative behavior can be defined as the study of the elements in a design process that can lead to the generation of an original solution, in particular the following aspects: new objects, activities performed, sequence of actions and models by which activities are completed (Snider et al. 2016). In the literature, an unusual starting point can lead to an unusual and original solution. But, at the same time, a creative result often requires the judgment of an observer (Amabile 1996; Csikszentmihalyi 1999). Prior to this judgment, the designer, their academic training, the intellectual process followed and the sociocultural context in which the work takes place can all play a role in the final form of the result. There are thus many parameters that affect their judgment of an action recognized as creative or not: creativity in one of the above-mentioned compartments does not guarantee the production of a result recognized as creative.

NOTE.– "Creators":

"Serendipity reinforces the share of the unprogrammed in scientific discovery and, with it, epistemological audacity" (Besnier 2013). The term "creator" has been adopted as a synonym for a new generation of social media spaces designed to support content producers in new ways. "Creators" can be paid by their individual viewers, who can buy subscriptions, send tips or fund new projects. The term "influencer" emphasizes a person's magnetic effect on followers, a nebulous charisma easily turned to marketing. "Creator, on the other hand, emphasizes that all those who publish on social networks produce something, getting involved in the collective effort to make user-generated platforms convincing and therefore profitable. This idea is marketable: the creators' economy has so far benefited from investment funding of 1.3 billion in 2021, almost three times the funding for the whole of 2020" (Les Actualités 2021). Do not confuse the real creator with the media creator!

Guilford (1956) showed that a creative process is based on two concepts: divergence and convergence. The designer exploring a range of possible solutions and information proposes a possible path and the designer will analyze the considered solutions through evaluation and combination to come up with a single appropriate result. In this scheme, the designer and conceptor can be the same person. On this subject, NAC (2014) speaks of research "inspired by a vision". Figure 1.2, inspired by this reference, schematically illustrates the convergence/ divergence couplings.

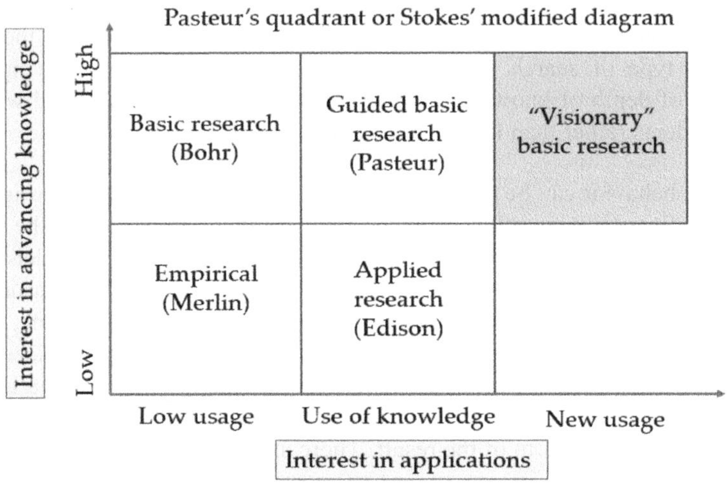

Figure 1.2. *Convergence and divergence associated with different areas of science*

Although both are essential in the creative process (Cropley 2006), as we will see in section 1.2, divergence in particular is characterized by exploration (trying to deviate from the norm in terms of solutions possible) of the problem posed and how it can be solved. This can be done either through active exploration of the solution space or through more passive deflection following single solution principles that deviate from those that are generally well defined or understood. This type of reflection is abundant in the literature and can occur through traditional and direct techniques such as those listed previously, as well as brainstorming (Osborn 1953) or analogy (Chan et al. 2011; Gonçalves et al. 2013), identification of emergent properties (Gero 1996), exploration of the problem through coevolution of problem and solution spaces (Maher and Poon 1996; Dorst and Cross 2001), monitoring of opportunistic design processes (Guindon 1990; Visser 1994; Bender and Blessing 2004) and reflection in the classic literature on creativity, as described by Wallas (1926) and explored by Boden (2004).

Valley of death no. 2: integrating a convergence/divergence system

NOTE.– Some relationship difficulties:

The convergence of heterogeneous knowledge can, in principle, broaden the combinatorial field to achieve a unifying objective, with ideas from disciplines other than those mastered by the initial bearer of the basic idea. Boardman and Bozeman (2006), Nooteboom et al. (2007), and Disis and Slattery (2010) show, however, that this collective breadth of vision can be a source of tension and misunderstanding between members. Cummings and Kiesler (2005) consider that each discipline has its own network (paradigm obliges) and the scientists of a discipline participate in publications and communications adapted to their knowledge. Under these conditions, despite their scientific relevance, they can be modest contributors to convergent and largely interdisciplinary activities due to the difficulty of working on interdependencies. Nooteboom et al. (2007) propose an inverted U to represent the relationship between "cognitive distance" and cooperation success: the optimal distance must balance the advantages of the diversity of knowledge and the obstacle to finding a field of knowledge, common cooperation. Moreover, when the spatial distance between partners increases, close ties become weaker, according to Cummings and Kiesler (2005). The new modes of communication via the Internet at least partially correct this effect.

Creativity therefore depends on the networking of individuals working together toward a common goal, but in principle without the constraints of processes and routines imposed by institutions (Parmentier 2020). It is, according to Gero (1998), a situated and temporal project of transformation, as it lies in the evolution of imaginations oriented toward technological concepts, then products. There is a twofold problem to be tackled here: on the one hand, recognizing the value of this kind of spontaneous, free process, and, on the other hand, finding an "administrative framework" to enable it to unfold under standard or crisis conditions.

Moreover, as Rias (2017) points out, new knowledge is absolutely necessary to differentiate ourselves from competitors and, at the same time, it is essential to quickly capitalize on knowledge that has emerged recently, in a framework where new knowledge must fit in as closely as possible with the technical culture of companies (another (useful) form of conservatism). In the United States, the federal government invests approximately $150 billion/year in basic research and R&D. What Link and Scott (2020) show is that technological changes could promote creativity in the production of scientific knowledge. They deduce the possibility of adjustments to the financial support policy that would have the potential to increase productivity in the scientific field. The disturbance of an environment would therefore exert an influence on creativity. Figure 1.3 shows multiple factors that come into play (Klein 2010), with obstacles and disincentives to interdisciplinarity, as well as strategies and mechanisms that facilitate convergence initiatives.

Figure 1.3. *Obstacles and disincentives to interdisciplinarity and mechanisms that facilitate convergence initiatives*

The achievement of an operational development can be informed by a multidisciplinary field of knowledge from people with disciplinary knowledge of varying levels. In the interests of rationalization it implies, we need to take into account the relationship between the knowledge called upon, and the space, time and people involved. "Individuals have preferences and motives for action (desires, needs); they know them perfectly, they seek to satisfy them intelligently within the constraints of their budget, they know absolutely the consequences of their actions once they have been carried out, and they therefore carry out these actions in the light of the consequences they anticipate" (Menger 2001). The (legitimate?) choice of construction partners, the resulting decision is made by a claimant, who must choose between future actors who can play a prominent role in proposing solutions likely to provide a framework for the idea. This implies a number of guidelines for deliberation and coordination economies.

Valley of death no. 3: an environment that supports convergence

NOTE.– Recommendations from NAC (2014):

On the subject of convergence, the U.S. academies propose the following recommendations:

1) Experts, funding agencies, foundations and other partners should identify key problems whose solution requires convergence approaches to catalyze new research directions and guide research priorities.

2) Research institutions, funding agencies, foundations and other partners need to address the obstacles of effective convergence as soon as they arise, notably by developing funding mechanisms for convergence efforts and supporting the review of proposals in collaboration with funding partners. Institutional programs such as seed funding to catalyze collaborations should be implemented or extended.

3) Institutions should review their administrative structures, faculty recruitment and promotion practices, cost-recovery models and research support policies, in order to identify and reduce the obstacles of the formation of inter- and intra-institutional partnerships that facilitate convergence.

4) Academic institutions should develop hiring and promotion policies that include explicit guidelines for recognizing the importance of convergent and disciplinary research, as well as criteria for evaluating them fairly.

5) Those wishing to encourage convergence should identify evidence-based practices that have facilitated convergence, drawing on the expertise of the economic, social and behavioral sciences, as well as program management and strategic planning. Understanding the unique barriers and strategies of the practice of convergence would improve practical guidance on how facilities can structure and maintain a convergence program.

6) Leaders and practitioners who have fostered a culture of convergence in their organizations and laboratories should develop partnerships, synergies and collaborations with their colleagues – particularly in smaller universities and institutions serving traditionally under-represented groups – to help institutions establish and nurture convergence efforts while promoting their own interests.

7) Best practices in effective technology transfer from research organizations to the private sector should be collected, established and disseminated. To ensure that convergent approaches enable innovation and stimulate future economic development, research progress must be translated into new products and services.

8) National coordination on convergence is needed to support the infra-structure for solving emerging problems that transcend traditional boundaries. Stakeholders in the convergence ecosystem – including agencies, foundations, academic and

industry leaders, clinicians and scientific practitioners – should collaborate to raise awareness of the role of convergence in advancing science and technology and stimulating innovation for the benefit of society.

To stay ahead of the game, at least in a troubled situation, it is necessary to go beyond the current increments which can already be significant disruptors (because it often happens faster than what most can accept, or the image of the frog in boiling water). Today, those who practice divergent thinking a little too much risk being eliminated from the group of researchers recognized by overly formatted evaluation structures. It is increasingly difficult for them to see their articles accepted and their credit applications are generally rejected because they do not fit into the dominant paradigms, as they are generally associated with risk taking. They do not actually burden the search system. The latter thus shows its lack of desire to take its own risks, a real form of conservatism which concerns the body of researchers as a whole (since as peers, they take part in evaluations), but also the system for managing research as a whole.

Taking into consideration what Baillargeon (2012) calls epistemic virtues, researchers must respect: intellectual honesty, integrity, the ability to submit what we put forward to the criticism of others, a certain and indispensable mistrust of our senses and memory, the ability to consider alternative hypotheticals, the practice of constructive doubt, the recognition of the fallible nature of our knowledge and so on. Are we able to produce ideas that are likely to lead to invention? For, as he points out, “it’s not enough to possess knowledge: you must also have the determination to use it, and the epistemic virtues that this implies”. This spark, the positioning of which is shown in Figure 1.4, must, after purely scientific reflection taking into account what is real and what is possible, move toward aspects that are oriented toward higher TRL and dissemination aspects (Hohberger 2016).

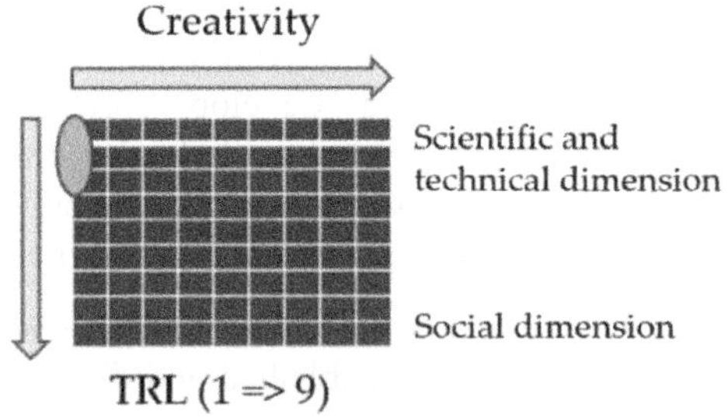

Figure 1.4. *Positioning the theme in the TRL/dimension*

The individuation processes that may be at the root of creativity simultaneously involve ethical decisions, wills and desires. Ethics is associated with the ego, while the other two elements are the guides of the unconscious. Jung (1964) considers ethics to be about unconscious processes. According to Simondon (1989), the perspective is based on an ontology, insofar as the internal processes between the "self" and "I" can be applied to connections between humans and technical objects, especially those with a high degree of complexity, concretization or concreteness. Creativity, or the generation of new ideas that are intended to be useful, is often considered in science and technology (Mumford and Gustafson 1988; Mumford et al. 1997; Mumford et al. 2005; Vandervert et al. 2007). Creativity comes into its own when solving complex, ill-defined problems (Besemer and O'Quin 1999; Ward et al. 1999; Ford 2000). For Serres (2013), "the misfortune came, in this philosophical path, from the foolish simplification of a question where baroque exuberance emerged. Simplification is generally achieved by means of a forced choice: continuous or discontinuous, analysis or synthesis, with the third party excluded. God or the devil, yes or no, with me or against me, of two things only one. Complexity beckons to the world of reality, whereas dualism calls for battle, where new thinking dies and the object disappears. Dualism is used to define niches in which fighters who lack courage settle in equilibrium for a long time to come", risking the death of creativity.

For Snider et al. (2016), creativity corresponds to a multidisciplinary field of research, with work ranging from psychology to architecture, digital sciences and the principles of engineering design. By relying on the imaginary, particularly to satisfy consumers, we need to understand or better anticipate their aspirations and therefore try to meet them in the best possible way. User sensitivity is associated with an advantage, a feeling of uniqueness or superiority, individual personalization, or the search for a response to an unresolved latent need, with the outline of possible innovations that will influence the future. Creativity is widely recognized as a complex and multifaceted research topic, and although it is often defined in terms of the creative product – original, appropriate and unexpected (Brown 2012) – it is important to consider the wider range of areas in which it can appear. To understand creativity, it may be advantageous to draw on the "pillars of creativity" defined by Rhodes (1961), which encompass the product, the creative person with their divergent ways of thinking (Helson and Pals 2000), the original, appropriate if unforeseen process (Lubart 2001; Sosa and Gero 2003) in which they exercise their talent (Amabile 1996; Lubart 1999; Csikszentmihalyi 2014). These four pillars are dynamically interdependent.

NAC (2014) reminds us that convergence with diversified teams is capable of producing innovative solutions to complex problems. Hong and Page (2004), Horowitz and Horowitz (2007), and Stahl et al. (2010) show that in these types of situations, the advantage goes to diverse teams over scientifically homogeneous groups. Thus, in a trusting environment where all opinions, however divergent, can be expressed, "where diversity is valued and opposing ideas are communicated with respect can be vital to the success of a convergence program. Such environments enable groups to think beyond entrenched paradigms and collaborate to discover creative solutions to difficult problems" (NAP 2014). On the theme of disciplinary diversity, Stahl et al. (2010) and Ely and Thomas (2001) show that disciplinary distance may initially be linked to difficulties integrating into an interdisciplinary group; it is possible to observe increased creativity and satisfaction after integration.

The traditional four-stage model by Wallas (1926) can be used to analyze the following elements: preparation, incubation, illumination and verification. In a creative proposal, it generally takes a long time for an idea to take shape semi-unconsciously. With the divergence/convergence model (1956), a process of exploration is required. More recently, Maher and de Silva Garza (2006) proposed iteration and re-evaluation as key elements in the formation of a potentially creative solution. The process followed by any designer is in turn governed by their own experience and decisions, the focus of the intended activity, traits such as personality, motivation, experience and numerous personal characteristics (education, culture, environment), the constraints imposed on the activity, as in complex design environments (Eckert et al. 2009), or design change environments (Eckert et al. 2012).

At this point, it seems useful to examine whether there are other factors that influence the development of creativity from family upbringing (Berry 1981; Clark and Rice 1982; Runco 2003), where it has been shown that people recognized as creative tend to come from backgrounds that value autonomy and intellectual activities such as motivation, accomplishment, awareness, autonomy, openness, flexibility and cognitive complexity (Mumford and Gustafson 1988). Moreover, exposure to multiple, complex and novel problems can enhance creative problem-solving skills (Mumford et al. 2000).

The aim is to highlight gaps or flaws in the current state of thinking (Lubart 2001), which can lead to a challenge to the status quo, a desire to solve an unthought-of problem, which can give rise to an idea. Kijkuit and van den Ende (2007) point out that the identification and structuring of the new problem is not always explicit, as the generation of an idea often takes place at the "margins of consciousness". Dunbar (1995), for his part, found that the laboratory in which scientists worked was a key factor contributing to the generation of new ideas, as we would be a permeable membrane to everything around us (Maffesoli 2000). In particular, when faced with undesirable events, they adopt coping strategies to manage their careers in an appropriate way, which in turn promotes their creativity (Tebes et al. 2004). Figure 1.5 recalls these couplings (Barrett et al. 2014).

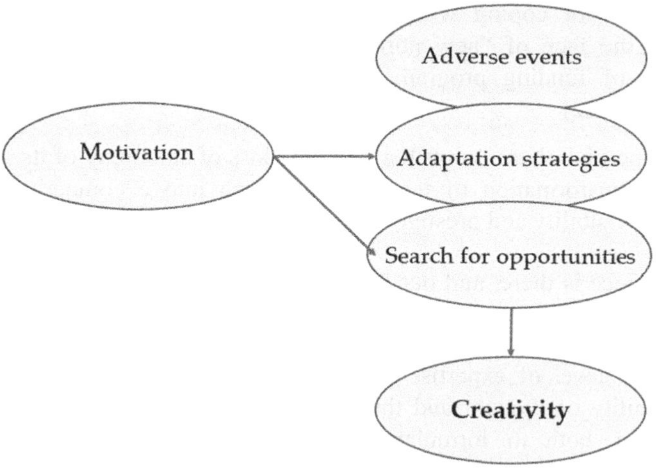

Figure 1.5. *Ways of promoting creativity*

Expertise, an essential component of domain knowledge, is essential to creativity in the sciences. With experience and a certain amount of scientific recognition, researchers can tend to refocus on problems considered relevant, even strategic, to pay attention to the obstacles to be overcome, etc. (Mumford et al. 2003; Root-Bernstein 2003). Several potential factors can be taken into consideration:

– Prestige: highly productive scientists sometimes tend to occupy high-powered positions (Allison and Long 1990; Koro-Ljungberg 2002). These positions, based on knowledge of the work of others, can lead to selection, resulting in a focus on cutting-edge subjects, with a display of creativity associated with a certain degree of autonomy.

– Outreach: this display can be a starting point for the development of a cooperative network (Louis et al. 1989). These connections can in turn help to generate new research ideas (Gruber and Wallace 1999; Koro-Ljungberg 2002). And so we come full circle.

– Serendipity: unexpected discoveries or their extension to discover the cause and verify its genericity (Dunbar 2000).

– Creativity without apparent cause.

– A certain reluctance on the part of the academic world for anything that runs (or could run) counter to the management methods of this body, despite incentives and policies to promote the value of research.

– Strategies for coping with "performance injunctions", the construction of meaning in the face of "new normative horizons" and their relationship to the mechanisms of funding programs often marked by confusion and uncertainty (Hubert et al. 2012).

– A competition that is contributing to the loss of autonomy of its scientific field and to the transformation of their own research into a competition focused on productivity, visibility and prestige (see the Shanghai ranking; Gonzalez 2014).

> Ignorance is there, and our knowledge moves in it like a tiny boat in the middle of the ocean (Jünger 2018).

> A high level of expertise acquired in a given field can reduce the flexibility of thought and the subject's ability to "step outside" this domain, both to formulate new questions and to elaborate new answers to the problems submitted to him (Bonnardel and Marmèche 2005).

> But knowledge is not about success. It is not a chain of clever inventions, an ingenious system for emptying crates and bags. This science will not save us. The Earth needs Man as a caretaker and shepherd. We must relearn to treat her like a mother (Jünger 2018).

NOTE.– Scientific productivity and performance:

Many question the constrained evolution of current modes of research operation, particularly those influenced by science programming, with significant resource allocations, administrative and evaluative reinforcement (Vilkas 2009; Barrier 2011). With these more quantitative approaches, indicators of excellence and/or performance have become commonplace, particularly the H-factor (Hirsch 2005). Gingras (2014) stresses the absence of epistemological foundations for procedures and their relevance, in line with other critics of New Public Management opposed to the quantitative alone (de Gaulejac 2012; Bourque 2016).

Do researchers who are highly productive in terms of the number of publications have a significantly higher probability of producing highly cited articles, and thus of being considered "prestigious"? Or is it a question of a multitude of irrelevant articles (decreasing marginal productivity)? The answer to these questions is always important, as it helps determine whether a researcher's influence can be judged. It is naturally imperfect, but, according to Sandstrom et al. (2016), quantity would make a difference. Is this a common misconception?

Valley of death no. 4: (still) having a creative idea despite a shaped education...

"The specialist, once his specialty has been transformed into a fortress, gives free rein to his will to power; under the pretext of division of labor, everyone wants to be master of their own home and defend their positions against external enemies and rivals from within". So wrote Gusdorf in 1968! And, to the best of the author's knowledge, this situation has not changed much, despite all the talk about opening up to other disciplines. According to Pasquinelli (2022), expressing our critical mind can reveal weaknesses in our mental capacities, due to numerous more or less blocking biases. But the recent (and future?) pandemic linked to the coronavirus has shown that, for part of the population, there has been adherence to unreliable explanations which, at least via social networks, have spread. Clearly, the learning process of scientific doubt can contribute to the development of a critical mind; it enables us to know what a model, a paradigm or a heuristic approach is. Trusting in the words of recognized experts in their fields (not necessarily those who take pleasure in appearing in the audiovisual media) can also be an element of reinforcement in the credibility of information.

However, the structure of teaching, with its focus on rapid learning of knowledge, resulting in an understanding of training as the teaching of "truths", is partly responsible for a situation where, at best, technical operations are practiced. "Teaching a 'discipline' consists of recommending standardized models. However, all consensual ideas suffer from weaknesses, contradictions and half-truths, which are always worth pondering. [...] Some ideas are consensual, not because they have been verified by the community of peers, but because their role is precisely to build consensus among peers and prepare the novices for a profession (a form of disciplinary socialization)" (Raynaud 2021). Simmel (1999) teaches us that:

> In social groups whose aims and meanings are as different as can be imagined, we find, in spite of everything, the same formal behaviors of individuals among themselves. Domination and subordination, competition, imitation, division of labor, party formation, representation, solidarity on the inside at the same time as closure on the outside: all this can be found in state society as well as in the religious community, in the gang of conspirators as well as in the grouping of economic interests, in the art school as well as in the family.

With this conservative lead, it is hard to be creative!

Some believe that a researcher's environment can/should have an impact on their creativity. If a researcher has been selected by their "faultless" academic and school curricula, by their performance in responding to standardized demands, we can expect an intelligent response from them. But will they be creative? There is a cognitive dimension associated with any intentional state, based on a body of knowledge about the possibility of accomplishing the action, in a given environment (Lewin 1951). "The most elementary teleological means-ends plan of action is based on what we know about the world in which the action is to take place. However, since two researchers in the same field never have exactly the same knowledge, this difference could be the source of variable actions or expectations of others' behavior" (Raynaud 2021).

But can we not reverse the question for two people who have received the same "supervised" training; could they not find the same solution? Such is the debate on the difference between a brilliant mind like that of Henri Poincaré and a free, unformatted mind, like that of Albert Einstein! Apart from the cases of these two extraordinary personalities, is there not a risk of a slavish use of the method and language of science on a given subject, without the creative spark being present? It is "a frankly unscientific attitude, in the proper sense of the term, since it involves a mechanical and uncritical application of habits of thought to fields different from those in which they were formed" (Hayek 1952). Thus, if understanding can be classified in a psychological category, and explaining in an epistemological category (Bunge 1997, 2007), can we separate this form from these contents, by means of scientific abstraction? So where is the creativity?

For those who are not part of the hierarchy of high-level scholars, segregation means being placed in a ghetto. Harroche (2021) writes in his thesis:

> While the reproduction and growth of inequalities are often seen as perverse effects of public institutions, many so-called neo-managerial reforms today aim to manage the public sector by creating, deepening or building on inequalities among professionals, organizations and users.

In France, the launch of the initiatives of excellence (IDEX) is just one example of the increasing competition between researchers, laboratories and academic establishments. Where are we going to place the imaginative ones in this setting that is not really made for them? This relegation refers to a mental process, and for Raynaud (2021), "from the point of view of social semiology, it is always preferable to replace an invisible process with observable clues. Stigmatization must not simply be associated 'in one way or another' with confinement; it must be the cause of ghettoization (and ghettoization must result directly from stigmatization)".

> But what is technology? [...] Increasingly, technology has become a kind of learned or pedantic superlative of technique. [...] The origin of

> this abuse is well known: it's the post-war habit of transposing the Anglo-American term *technology* into French, in ignorance of the meaning it already had in our language (Sigaut 1988).

> We will examine with the utmost attention the true essence of the "intentionality" that Nature has placed in each of us to varying degrees. For the very concept of intentionality means that we are more concerned with the results of our most distant actions in time than with their intrinsic quality or their immediate effects on what surrounds us (Keynes 2021).

Furthermore, knowledge of the scientific field can (must) give rise to new projects, enabling the researcher to explore other original avenues. Root-Bernstein et al. (1995) found that, for top scientists, changes in their field of excellence were necessary for long-term productivity. "Because creative work tends to be very intense in terms of cognitive resources, the shifts in work and mixing of projects will perhaps allow scientists to manage their cognitive resources more efficiently, thus allowing them more opportunity to engage in creative thought" (Barrett et al. 2014, p. 49).

According to the same source, prestige is detrimental to technical influence and qualitative reasoning (while many scientists wish to occupy positions of power). Fame is said to be a long process, whereas creative work, which is part of immediacy, can take place throughout a person's life. With a decision-making position, responsibilities can change and take a detour from creative work. From these observations, creativity, if it exists at all, needs to be recognized as such if it is to have any chance of being transformed, either into original scientific activities (with the risk of upsetting existing paradigms), or into useful future technical objects requiring a decision to be made (sometimes with rejection of the operation, as shown by Quan et al. (2020)). If the researcher is not very well known, this transition is unlikely, on average, and if, on the contrary, the researcher is very well known, the risk is that they will not be very creative, having become a generalist! Before we think about POC, the necessary demonstration of an idea, in this context, reveals a new "valley of death" for the promotion of that idea.

Valley of death no. 5: having a creative idea that is recognized as such

1.1.3. *Methods for developing creativity*

> Imagination consists of imagining absent objects and combining images of these objects with one another. It virtually brings new possibilities into existence. In this sense, it frees us from frameworks

> and leads us to see the world from a different angle, breaking away from the dominant logic. It allows us to experiment with a wide range of possibilities, leading to new ideas (Parmentier 2020).
>
> Innovation is made up of two main phases. The upstream phase is fuzzy and relies on creativity while the downstream phase is swift, relying on discipline to get to market (Rias 2017).
>
> What decision-making mechanism would allow all citizens to express their point of view, not on research itself, but on the questions raised by the applications of research? (Heidegger 1958).

In a free interrogation stage, according to Krichen and Chabaud (2019), idea-generating proposals can be put forward by a group of people with the aim of proposing a solution to a particular problem, before outlining a potential contextualization with the invention. The first stage involves a variety of free ideas. The more numerous the proposals, the wider the choice can be for a subsequent operation. In the second stage, the proposed ideas are evaluated in order to select the one with the greatest potential for added value. Identifying such an opportunity involves finding a solution to an unmet demand or need. While the initial brainstorming is easy to implement, the activities of filtering and mastering the links between scientific and technological knowledge and a possible business opportunity require specific knowledge, skills and aptitudes.

Creativity is an integral part of human activity and can, if successful in terms of innovation (TRL > 6), be a factor of personal and organizational success. But this is not always true: let us recall the fate of Nikola Tesla, a particularly creative engineer who was largely forgotten by history, and more or less exploited by George Westinghouse. Even so, his name stuck to a magnetic field measurement unit! There are many different approaches to supporting creativity. For example, the process of creative scientific discovery is generally modeled on C-K design theory (Shai et al. 2013). Idea generation can take advantage of numerous methods (such as TRIZ, C-K, six Bono hats, brainstorming, mapping, etc.) (Doran and Ryan 2017). For its part, the Delphi method integrates intuition and expert knowledge in an attempt to "know the future" (Kin et al. 2021) by identifying emerging trends as accurately as possible, while discussing hypotheses for the evolution of our environment (Rousseau 1996). According to Parmentier et al. (2021), it remains difficult to assess their effects on innovation with any certainty. The idea may be generated spontaneously to solve problems (those remembered) (Unsworth 2001).

Western states support the idea of the "researcher's freedom" who can, in principle, be creative. According to Mehrotra et al. (2021), in the corporate world, it would seem that employees are more constrained, although a demand exists.

Companies find it difficult to set up creative environments, not least because of the subtle processes that hinder this culture: slowness of decision-making, distribution and number of decisions, speed of decisions (premature project closure), destructive pitfalls of critical discourse on the project (sleeve effects), mastery of brainstorming management (group or "nominal"), etc.

1.1.4. *Estimating creativity*

In an in-depth evaluation carried out by peers on behalf of clients, is the proposal made by one researcher or a small group of researchers a matter of appreciating what has been understood about preferences, or of satisfying material constraints? Do these points of view provide antagonistic or complementary perspectives? These are questions already posed in Demoly and André (2022). But it is important to avoid a totally "creaticide" operation on the part of our own assessment and that of the responsible decision-maker, which would risk any initiative being "nipped in the bud". In that case, we may as well do nothing, or go back to the way things were before! To put it simply, six criteria have been selected and presented in Figure 1.6, criteria of opportunity in relation to the spirit of a disruptive project, but also of feasibility.

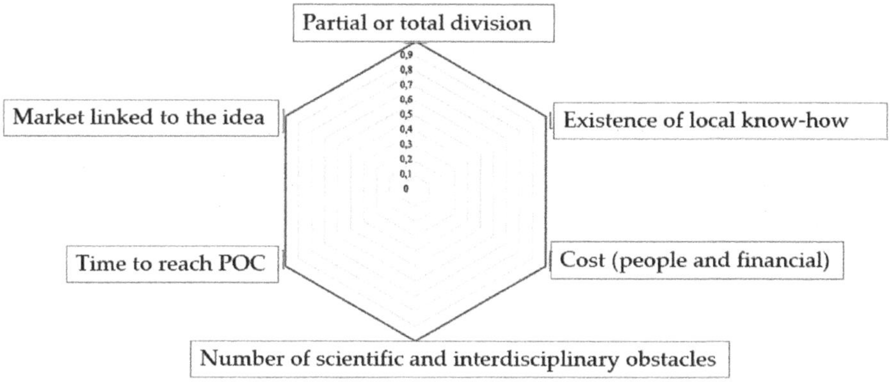

Figure 1.6. *"Strategic" evaluation basis of a project (POC)*

The idea is to represent a (very approximate) digitization of each of the criteria presented in this figure by a polygon, and examine whether the center of gravity of the polygon thus created is located toward the top of the hexagon (favorable position) or, on the contrary, toward the bottom (questionable feasibility). To indicate the low degree of finesse retained, a duration is low (e.g. less than one month), reasonable (six months to one year) and high if it exceeds several years. It is up to the experts in a group of researchers and/or evaluators to adjust to each other based on a learning process, which depends on the "means" deployed by the organization on which they depend. Depending on its financial strength and pioneering spirit, the criteria may naturally evolve.

To illustrate the applicability of this presentation, we have chosen the early example of an innovative 3D printing project, launched in 1984 (see Figure 1.7). At the time, it represented a real breakthrough. With the contribution of adaptive materials (available "off the shelf"), the same graph could be proposed today for 4D printing (Demoly and André 2022).

Local know-how was present in 1984 (processes, chemistry, polymers and automatics) with a motivated person, the initial market was clear (at the time, it was primarily a question of producing prototype parts using a 3D photocopier), the interdisciplinarity was managed by a single person, which enabled a POC in a few days by borrowing resins from the neighboring macromolecular chemistry-physics laboratory. Clearly, the polygon's center of gravity is well located upwards and falls within the criteria presented above (André 2017b). Nevertheless, and this is less obvious, it is still possible to ask whether the proposition would be supported by scientific experts today (assuming that additive manufacturing has not been developed)? The example of 3D printing will be discussed in the second part (Chapters 2 and 3 of Volume 2) of this two-volume work.

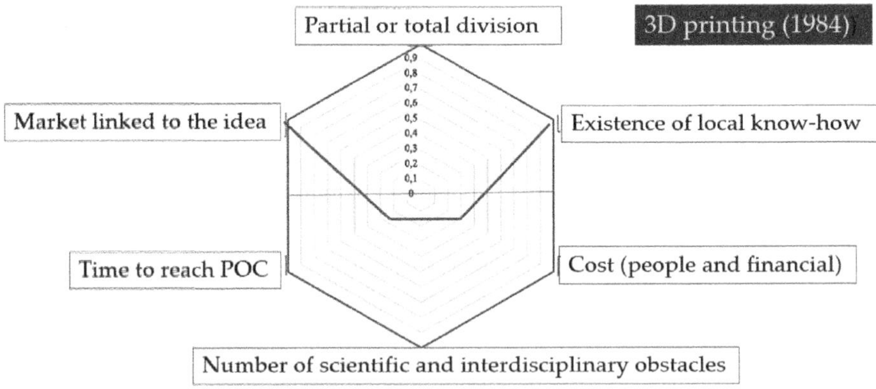

Figure 1.7. *Example of 3D printing (applicable to 4D printing)*

NOTE.– Dunning–Kruger effect:

Kruger and Dunning (1999) showed that participants who performed less well on humor, reasoning and grammar tests were also more likely to overestimate their performance. Overconfident people are therefore low performers, both "unskilled and unconscious". Jansen et al. (2021) complete this explanation. This bias effect can be found in many surveys and in the emergence of ideas.

1.2. Proof of concept

> We talk a lot about understanding and communicating [...]. We believe there are great secrets to be discovered. We hope that Science will show us the way [...]. We dream of a life in the distant future that will be radically different from the one we know today; we ascribe to ourselves indescribable powers (Miller 1957).
>
> The self can be seen, if you like, as an inner god, because it pushes us toward wholeness, our totality, the integration of ourselves with our own shadow, our wills, motivations and projections, and the authentic recognition of who we are and what is really important for our individual history (Pinheiro Neves 2011).
>
> He had studied to be a surveyor, and the positive sciences had influenced his mind to the point of making him a practical man without much imagination (Ray 1967).

Valley of death no. 5 bis: having a creative idea and integrating it into a rational approach

Le Chatelier (1925) organized scientific research on the model of an industrial enterprise. He envisaged a cost/benefit of scientific production in terms of the inventions it could produce for national industry, with a certain obligation to achieve results. It was against this model that Jean Perrin developed the concept of "pure science", because science must remain a creative, disinterested activity with its own internal purpose (obligation of means).

This section looks at how we can move from a (supposedly happy) idea to an application (which should be just as happy) through a scientific and technological demonstrator: the POC. A question common to the design of products and/or objects concerns the transmission and material translation of an idea.

This is a crucial question, because an idea, usually an individual one, must then be shared by several people, in different professions, between demonstration (POC) and, later on, industrial development, which will use its own design methods. “It [the idea] establishes common positions and orientations for the feasible and the representable, and thereby holds together, in advance and by construction if you like, the crowd...” (Castoriadis 1975).

1.2.1. *Integrating an idea into a (hopefully) rational framework*

> We cannot choose but admit the right of the citizen to adopt a condition of voluntary outlawry (Spencer 2010).
>
> As such, [mental representations] represent specific characteristics in terms of the organization of the contents of mental operations and logic (Moscovici 1976).
>
> Only a few artists, writers and scholars escape the cold shower of normalization, the return to order, the loss of the magic of the time when everything was still possible [...]. It is in this that all art, all innovative thought, is transgression (McDougall 1978).

The function of ideology is “of reconstituting on an imaginary level a relatively coherent discourse which serves as the horizon of agents’ experience” (Poulantzas 1968). At the start, we can have a vision of an original action without the support elements being well defined. The person who carries the idea has no reason to be a Cartesian subject “guided exclusively by the lights of the reason” (Giust-Desprairies 2003)!

In general, the first step is for the person who has created this imagination. They will try to structure their thinking so as to envisage the realization of the POC. If there are too many obstacles to overcome, or if the actions to be taken are too far removed from their body of knowledge, the idea risks being self-censored and ending up in the garbage can. At this stage, some ideas are “simple” enough for a single person to draw up functional specifications for the POC. Others may, through a pragmatic apprehension of reality, gain in robustness through exchange with partners from other backgrounds.

They must be known to the applicant and willing to play the interdisciplinary game. This is not always easy if they are not (for example) partners in the POC project. In a group such as this, with all the voices present, a binding force needs to be created. Because everyone has found an interest in the idea, a principle of order must be expressed, translated into a form of potential investment in an imaginary that has become collective. It is worth remembering that this operation is normally free of the institutional system (social regulation, transmission of values, codifications, etc.). However, the latter's mission is to order and structure ideas around the paradigms that normally govern scientific practices. Disruptive ideas, promoting a new artifact through the interdisciplinarity they may generate, may break out of this normative framework and come into conflict or friction with established, tried-and-tested structures. However, they are based on the classical foundations of the "previous world".

In this pre-POC approach, we need to identify building blocks that are representative of a POC project, with the potential to evolve from a scientific objectification process (metaphorical richness, agility, flexibility, interpersonal logic, deconstruction of presuppositions, scenarios, etc.). By playing with different partners, the associated top-down approach imposes a teleological approach (governed by finality), which must be translated into effectively interdisciplinary workings. At this stage, the people involved in the design process exchange information, in the form of representations, with a view to communicating with each other for the better realization of a project which, at this stage, has no physical existence whatsoever.

While this situation is risky in terms of results, it is positive in that it places all the partners from different backgrounds on the same level, and connects scientific silos that are still too disjointed. For this exchange to take place, scientific popularization work is required, so that each partner understands how to rationally organize a pre-POC, with interdisciplinarity drawn (alas) downwards (because it has to satisfy all the members of the group):

> Popularization ensures the reflexivity of science, a reflexivity we need to come back to what we're doing, to our object of study, through another set of eyes. There's something fundamental about the spoken word in popularization: it's reflexive by nature, so it's a good tool. The researcher oralises his science. He passes it on to the public. To do this, he needs the other: the public, the non-expert (Jurdant 2011).

However, what is expected is a functional interdisciplinarity which must merge various disciplinary points of view to build a sort of theoretical mesh of the complex problem. It is then translated into a proposal located in a reference code that is as familiar as possible and therefore understandable (with its benchmarks, its locks and its explanation systems). In fact, these priming proposals are part of the imagination and are subjective constructions by association of personal and collective images, and naturally by relational artifacts. This last area, which is an element of very particular sensitivity of the pre-POC, can cause the operation to come out at the top (toward the POC) or, on the contrary, result in destructive phenomena. For example, "even in a group freed from power issues, another pitfall could arise, that of implicit information".

A rough draft corresponds to the construction of a mental representation at the interface of disciplines. Each member of the think-tank uses initial observations to identify a range of possible processes. This is a guide for characterizing initial conditions and building the tools needed for observation. In a convergent approach, it is necessary, according to Bousset (2012), within a framework of acceptable confidence (Bourdin 2022), to co-construct a mental representation: "Through exchange – provided it actually takes place – in the course of a gradual process in which the partners pool not the description of observations or the results of hypothesis tests, but links between ideas, which they gradually and reciprocally consolidate".

Remember that a group, even an informal one, is a social structure, with dress effects, always in motion, animated by impulses and complex interpersonal relationships ("always more or less the prisoner of mutual fascinations" (Lévy 1997)), a structure that can create order or, on the contrary, disorder. What is expected of this group is the construction of a collective reality defined as an instrument for orienting perception, at a given moment, and for elaborating a reasonable collective project. Indeed, we need to continue to give meaning to the action (recognized as such for the rest of the operation) in order to envisage a useful investment. Thus, according to Bousset (2012), the principle construction of the pre-POC could be carried out according to several registers (see also Magni et al. (2009)):

– production for each concept, a diagram that can be understood by other disciplines;

– decision to select and link together the concepts relevant to the proposition to be dealt with;

– the adjustment of concepts to each other in the "context of the whole", while preserving their reference to their original discipline;

– once the framework has been accepted, it is broken down into task proposals for the POC's realization.

NOTE.– Practical expression of representations:

For Boulekbache-Mazouz (2014), the representations associated with a possible shaping of the idea can be mathematical, graphic and/or textual. They constitute storage, cognitive artifacts to ensure a mediating function. "In recent years, new work techniques have emerged (project platforms, parallel design, collaborative work, etc.). The aim of this new way of writing is to activate shorter, more efficient processes by parallelizing or simultaneousizing tasks, resulting in better quality, less expensive works, integrating all players in the production chain as early as possible in the writing process".

Leleu-Merviel and Boulekbache-Mazouz (2014) define, in their work, forms of standardization that promote interdisciplinary exchanges with contributions in the following areas:

> – "Can we locate the designer's thinking through the synthetic form of formulation that is representation?
>
> – How do we conceptualize? And how do we represent what we have conceptualized and/or designed?
>
> – Do formal project representation tools influence or condition thinking?
>
> – Do design representations promote exchange and standardization between players?
>
> – Do certain tools provide a better understanding of design intentions?
>
> – Is it possible to define a model of design activity? And if so, how?"

The question then becomes who decides what; for example, Holahan et al. (2014) use a formal process to select divergent ideas, while others propose specific selection criteria for each type of idea (Magnusson et al. 2016). Ininou and Loilier (2021) give an idea of the many questions involved in structuring a creative idea.

Other actors, "interactors" who are no doubt apparently "less glorious" than the imaginative or the great engineer, can "translate" the knowledge coming from the idea and turn it into reality. However, these intermediaries need to master the languages and concepts of those they will be "translating":

> The sciences can no longer content themselves with accumulating huge masses of scattered data that are difficult to exchange, because the lexiques of the disciplines that are constantly becoming more specialized have become foreign languages for anyone outside the scholarly community that elaborates them. This is not to mention their increasing inaudibility to citizens who see them only as techno-scientific results, praised when they are perceived as so many Promethean promises, or reviled when they are accused of leading humanity to its doom (Silberstein 2013).

Are the same types of scientific personalities not involved? Or how, on "cutting-edge" subjects, can we have the right skills, especially when a disciplinary language barrier can limit the transfer of knowledge? This discipline-specific language is primarily designed to save time between members of the same community; a group factor in the recognition of belonging, it only imperfectly enables interdisciplinary exchange (Saint-Sernin 1979).

This observation is reinterpreted by Henry (1998), who writes: "It is one thing to observe that science solves problems; it is quite another to believe that science serves to solve problems. Moreover, does leaving our own discipline lead to a faster scientific career (Vinck 1995)? What will the members of my scientific community think? Here too, conservatism reigns supreme. Are the disciplines of science today not reminiscent of the old craft guilds destroyed a few centuries ago by incipient capitalism?

> In fact, scientific work is more akin to that of the handyman or craftsman than to that of the technician or engineer. [...] The scientist is described as a do-it-yourselfer [...] who possesses know-how. They use their know-how opportunistically, according to need and experience [...]. From this point of view, the laboratory is seen as a resource center for scientists (Vinck 1995).

In truth, at this stage a question emerges about the true value of this activity (which can/must serve as an example for other ideas), unstructured, calling on good will, if possible, with a high level of scientific and technological expertise. Is it possible to build an ethic of this essential work that is satisfactory, both for the person who is at the origin of the idea (if this author, at least, does not want to satisfy the "fragile pleasures of pure vanity" without seeking to go further than the enunciation of the idea) and a "clientele" of readers, that is to say, which satisfies forms of egotism, but not altruism? From this informal exercise of creativity resulting from a group formed in a somewhat random way (which we push to reflect or to reason in a combative, interactive way to produce a plan of action), can we obtain, at the same time, a self-advancement, knowledge and control of our own intellectual faculties, stimulated by the exchange between members, and moreover, a legitimate, powerful and fruitful modification of minds, associated with the creation of a collective imagination, bearer and support of an invention? How can we prove that the scientist does not "work in their corner" and that their research work is intended to be offered on a form of open "market" so that, at least someone takes it and uses it? Finally, participating with others in this context is a form of recognition of our own activity:

> At the very heart of the thought of the scholar or artist most absorbed in his research, and who seems most entrenched in his own sphere, face to face with what is most self and most impersonal, there exists some presentiment of the external reactions that the work in progress will provoke: Man is hardly alone (Valéry 1957).

1.2.2. *A transition to POC*

> As long as a product or service is not on the market, everything remains "imaginary": the relevant offer, the real targets it will reach, the uses that will develop, the business that it will generate, the actions of the competition, often even the technologies that produce the innovations (Musso et al. 2007).
>
> We need to keep an eye on what is emerging, but there is no reason to forget resilience [...], the weight of rigidities and friction. It's important to maintain the process of hybridization, diversity and irreduction, but also to think about perpetuated vulnerabilities and impossibilities of doing things, to analyze domination for itself, the way it is implemented and the way it is experienced (Pestre 2014).
>
> If researchers go so far as to confront this question of use (which is not their job), when this question is known to be particularly difficult by all those who are interested in the design of tools, it is good that the return sent back by usage, therefore the confrontation to practices, constitutes the only possible returns capable at a given moment of calling into question, in a well-founded way, the theoretical advances on which they were built. Beyond the scientific refutation, the test of use is the encounter with a socially, humanly and technically situated reality (Aggeri et al. 2009).

In the previous operation, the members of the informal group play a very particular role, as applied to possible practical needs. But, without having yet reached the goal of the POC, they are a means of an evolution of which they do not know if they will be able to reach the final direction and its end. They do not know the destiny assigned to their work (which they support) of which they have integrated the exciting part which is directly sensitive, which responds to their desire and their goal. They do not yet see (?) the following repercussions (unless they are integrated into the POC production group). From a practical point of view, it is therefore necessary to engage in the search for synergies between disciplinary fields, research which conditions the object of the action and its relations between science, technology and society (Ollivier 2002), "not entirely foreseeable by knowledge of the prior art", as claimed in the patentability of innovative works. It would also aim, for (Gramaccia 2008) "the social integration of knowledge for more social and cultural mediations". However, if it is already necessary to select the few interdisciplinary actors, if we do not seek to "recognize" our own deviant "children", it will never be possible to define the concept of interdisciplinarity more precisely, nor to exploit it.

In the technical field, which is broader from a disciplinary point of view than that of the scientist, knowledge is even more incomplete and provisional. What is more, the evolutionary process is accelerating. "In the absence of a 'belief' that is unanimously and explicitly shared, and whose content is held to be true, we strive to maintain the credibility that is satisfied with the plausible" (Perrot et al. 1992).

This situation is becoming increasingly fragile, and the many cracks it creates are leading to misunderstandings and demands for explanations on the part of society as a whole. In this context, the fragility of the individual in their environment is a feeling inspired by the idea that this environment is "inhuman". The environment is what Roqueplo calls "techno-nature":

> Our "natural" environment is almost entirely the product of technical activity. In this sense, our 'nature' constitutes an immense technical object. It is this object that I will refer to as techno-nature (Roqueplo 1983).

One condition is to fully understand the scientific information required by the requester, before providing knowledge. Clearly, this request is generally not well defined in its initial phase and will evolve over the course of discussions between the various members of the group. As far as possible, the initiator would like to have access to decision-making tools that can be used on a personalized basis, rather than as an application. It is a good idea to offer a tool that incorporates some of the complexity of real-life situations understood by the members (such as the existence of multiple criteria, allowing trade-offs to be considered), which enables an understanding of the processes and self-learning by the entire group.

To this end, the involvement of actors in the definition of objectives and the organization of exchanges between the various players via the construction of social networks can be useful. During this operation, a number of trade-offs must also be made between contradictory objectives such as tactical approach (monitoring) versus strategic approach, strategic exploration of the different options versus accuracy of forecasts, scientific production versus problem solving and observer posture of researchers versus facilitator posture.

The analysis by Sriskandarajah et al. (2006) is that the quality of relations between research and development and actors is generally not sufficient to allow the involvement of all actors in a constructive dialogue. They recommend a greater lucidity on the objective of the pre-POC, namely the relativity of the relationship between the goal and the social need and the representations that we have of it, as well as the limits of the "toolbox" which we dispose of:

> Thus, the angle of attack of a problem and the type of scientific approach to deal with it will be positioned in a gradient – not always well identified – between a reductionist and a holistic point of view. They should also depend on the characterization of the problem on the double scale of the degree of consensus and the degree of technical complexity of the problem identified (de Turckheim et al. 2009).

Gathering knowledge on the functioning of a system and representing it in a formalized way, generally in the form of models, is a classic scientific approach in research because its objective is to give a shared representation of the problem and the knowledge available, whether scientific or practical. This is the basis on which problem solving and the design of the tools that will eventually be developed are built. If the models of piloted systems make it possible to introduce technical variables as input variables, the decision-making processes are not only limited to these acts, even located in time. The formalization of a decision problem must therefore integrate, to varying degrees, physical aspects for technological purposes and decision-making aspects.

NOTE.– Decision support:

According to de Turckheim et al. (2009), a decision-making tool is made up of three elements: scientific or technical knowledge, an action objective and a definition of how the artifact is to be used (at least in the first instance). The pre-POC approach should include a vision of future use in relation to the present. This work, which involves all three points of view, aims to integrate tools into technical-economic engineering and can already incorporate potential uses (Hatchuel and Weil 1992). It is a way of examining how the idea can lead to the transformation of a situation (change process). "The use of the tool will be as much an opportunity for learning as the design of the tool itself. The scientist's and the future user's points of view therefore need to cooperate when designing a new tool and its use".

According to Encyclopaedia Universalis (1995), "innovation results from any change consciously introduced into the economy by any agent with the aim and result of using resources more efficiently and satisfactorily". In line with Musso et al. (2007), our "approach to innovation is neither romantic nor rational, but integrative". It is based on intentionality and on systemic or holistic approaches, drawing as far as possible on experience and, when several players are involved, on a cross-fertilization of perspectives, which can itself be creative. On the basis of favorable relationships, the agility allowed should make it possible to overcome the obstacles generally associated with interdisciplinary approaches. Sfez (2002), to avoid a disconnect between invention (rupture) and "servant" techniques, introduced the concept of "innovention". The idea of co-invention, as it emerges from the preparation for the realization of a POC, underlines its shared/mutualized aspect between multiple actors (scientists, designers, even perhaps manufacturers, mediators, users, etc.).

Przyluski (1942) completes these considerations by recalling that innovation does not only concern innate or even hereditary contributions, but habits rooted in a group culture. He writes: "It would even be an understatement to say that this [context] is limited to the past, because, to a large extent, the behavior of a being is determined by the anticipation of its needs. The social aspirations or what can become so fit well into the process". In this sense, innovation clearly involves decompartmentalization in order to achieve a goal (see also Leroy-Gourhan (1973), Schumpeter (1942), Gille (1978) and Alter (2001)). Thus, Koestler (1964, p. 104) writes: "The act of discovery has a disruptive and a constructive aspect. It must disrupt rigid patterns of mental organization to achieve the new synthesis". It is therefore a path of rupture of inertial processes, which normally leads us to believe that tomorrow will enter into continuity with the past. In any case, the invention is based, whenever possible, on the principles of economy (or optimization), contingency and convergence by:

– reflecting (watchfulness, imagination) on the possible assimilation by different users (from research to the public) of the idea;

– envisaging the integration of the new element into an existing set (only one variable is varied at a time);

– respecting principles of financial reality (the cost must not be prohibitive, making possible applicability unrealistic);

– considering the transfer of the idea to other niches.

These various principles, already anticipated in other contexts by Leroi-Gourhan (1973), are based on the following elements: assimilation, plurality, alignment and transfer. They are still applied today by cultural continuity in industry-oriented research (Guzzetti 1998). Technological evolution clings to pre-existing knowledge and practices in the POC before the act of innovation or demonstration of the idea's commercial feasibility. The result is a slow, painstaking process of exploring the full potential of scientific rationalism and technological progress.

However, we must be able to rise above previous facts, to group them, classify them and draw developments from them, to suggest rational modes of action linked to the disruptive idea: to make reasonable that which is normally not, which corresponds to an activity that may be necessary to convince and go further in the overall innovation process. Alter (2002), however, rightly makes a distinction between invention and innovation:

> [It] is heuristic because it allows us to understand that only a particular social process, appropriation, allows us to move from a discovery, an idea or a new product [invention] to new social practices that make it possible to integrate the novelty into the social and economic fabric [innovation].

In the technological field, it could no doubt be preferred to consider only the technical aspect, but for reasons of social responsibility of research, this leads to the analysis of the entire space between invention and innovation to varying degrees according to the imaginaries studied.

It was recalled that according to the classical vision of creativity (founded by Guilford (1950, 1956, 1959, 1967)) on the dichotomous principle of divergence/convergence, the creative process begins with the recognition or creation of a problem. From there, a process of divergence begins, culminating in convergence (see Figure 1.8), in a new solution to the problem (when it exists). So where does this new buzzword – disruption – fit in? It is the ability of the creative or the divergent to bring together seemingly distinct worlds, to find links where none existed before, to look outside a scientific discipline, a profession or a specialization to be open to everything, without taboos.

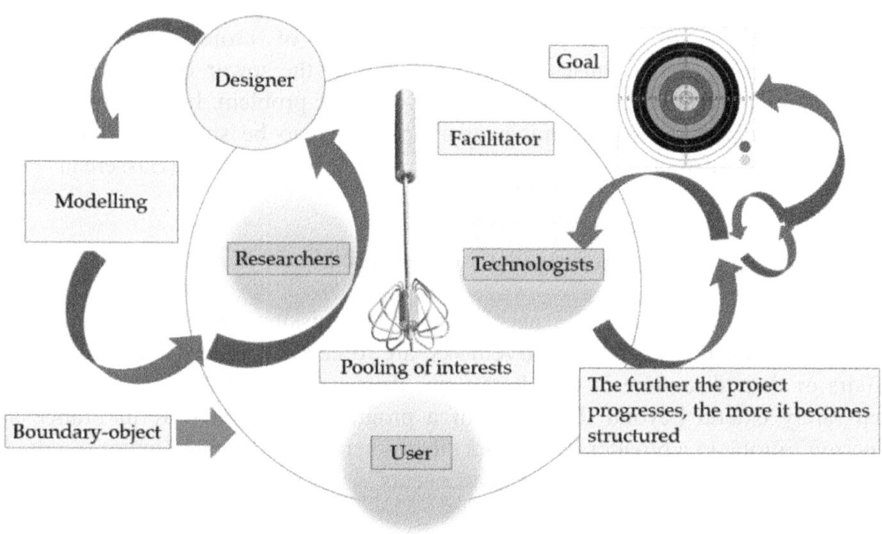

Figure 1.8. *Convergence players*

In a way, it is a kind of "adventure" thinking. It invites us to confront/invent the new and the unexpected. This relationship of transformation, made up of bricolage, adapting to circumstances and the unexpected (serendipity), (voluntary?) oblivion of pre-established ingenuity, requires us to invent a specific mode of transmission, one that is not based on the teaching of abstract knowledge. However, achieving acceptable robustness of this technology has yet to be achieved, hence the idea of a demonstrator. Then, a concrete but unstabilized object is a system in which interdependencies are exercised according to all (if indispensable) the laws of science.

We need to find ways of developing sharing spaces that are interdisciplinary, heuristic, temporary and open to research, enabling a more collective approach to creating and exploiting new concepts. In this way, discontinuities can emerge, leading to scientific recompositions and adaptations. That is why we need to account for the way POC-type objects return to the social world. It is a central question, which is beginning to engage research toward application, i.e. economic, social and political application, and puts this realization, in a responsible way, in direct connection with society. And yet:

> The context of simultaneous engineering and project platforms encourages the coming together of different design players, who never speak the same language. Worse still, they use similar words with different meanings in different professions and disciplines. This is one of the first paradoxes of simultaneous engineering, [the Babel paradox (Bassereau et al. 2015) represents a risk of project deadlock, analogous to the causes of the stoppage of the tower of the same name], born of a partial response to a complex problem. It arises from *design* projects which, however simple, need to be studied in their diverse attributes through an interdisciplinary approach (Bassereau et al. 2015).

The question of sharing, particularly across disciplines, is critical.

Increasingly, however, as a necessity of their mission/profession, scientists, in addition to their many administrative tasks, are forced to consider and deal with the affairs of their discipline in the short term; they are obliged to envisage durations that rarely extend beyond the scope of a program or plan (ANR in France and Horizon Europe in a broader framework) that is sensitive to events (major trends, for example): in short, they become “men of events”.

Generally speaking, however, the development of relations and contacts between the parties involved is becoming so complex and of such considerable dimensions that the mergers sought run the risk of becoming, or simply being, illusory. The issues to be dealt with quickly become drowned or disfigured in the difficult integration of disjointed areas of knowledge. By committing ourselves today to ideas that are too complex to deal with, we can rigorously consider that we no longer know what we are doing. This is the argument in favor of the "KIS" (Keep It Simple) program, which places the transcription of an idea into a "human-sized" proposal (and, as we shall show below, it is not that simple to turn an idea into an innovation).

Aside from the scientific and technical aspects, Giust-Desprairies (2003) insists on what is unknowingly and dynamically organized in the group in the form of a unit, a central imaginary meaning, "but which presents itself as an ordering principle, a binding force determining group functioning". This active, meaning-producing construction (with a status of objectivity) involves the group itself, the environment, objects and their attributes, and the pre-existing relationships between these objects and their attributes. "The institutionalization of shared illusion gives a sense of control" (Giust-Desprairies 2003).

In fact, the final object intended for users must go through an intermediary process of trial and error (Leroi-Gourhan 1973). This observation, which is almost self-evident to the "man in the know", leads to the assumption of time-consuming practical follow-up steps, which are essential if we are to consider the possibility of industrialization after an exemplary POC. These are deliberate, intentional representations of situations in which the players intervene in the classic register: seeking to model in order to understand and to understand in order to do (Le Moigne 2013). However, we all know that it is not enough to list the disciplines and professions required for design in order to put them to work (Minel 2003; Carlile 2004; Hatchuel 2006). From a creative proposal emerges the need for a POC, the positioning of which is shown in Figure 1.9.

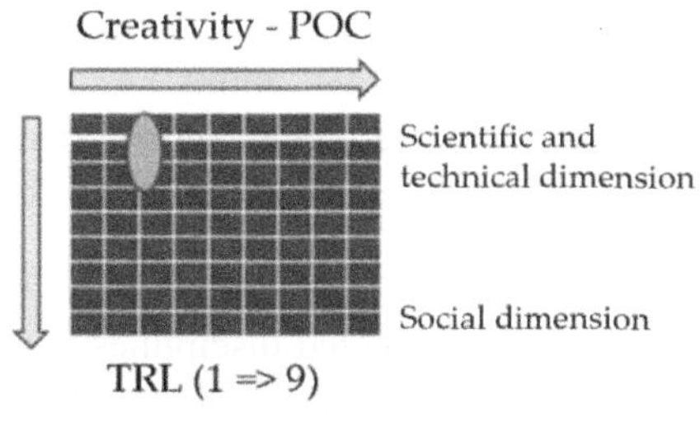

Figure 1.9. *Positioning the theme in the TRL/dimension*

1.2.3. *Basis for building a POC*

> The question then becomes how, within the laboratory, for example, the articulation between content and context takes place, in other words, between scientific practices and social practices. (Vinck 1995)

> Thanks to myth [...], ideas of reality, value and transcendence slowly emerge. Thanks to myth, the world can be grasped as a perfectly articulated, intelligible and meaningful Cosmos (Eliade 1963).

> We take care of other people's opinions and knowledge of others, and then that is it. We have to make them our own. We seem properly like those who, needing fire, would go to their neighbor's to get some, and having found there a beautiful and large one, would stop there to warm themselves, without remembering to bring any home (Montaigne 2020).

The complexity impacts the formal and functional characteristics of the demonstrators through interdisciplinary obligations. So, even if the translation of the idea is simplified so that it is, despite everything, transposable into reality, what is expected in this phase is to consider efficiency as a value in itself, the POC as a finality, relying on a rationality reduced to causal relations (radical treatment of forms of complexity) and therefore manageable. Leaving the classic models, the group associated with the pre-POC (like its successor, possibly identical) must be able to face the uncertain, the unknown as an element or as a stimulus of the creative process. "What is true is what works (instrumental rationality), and that it works is an end" (Giust-Desprairies 2003).

Malecki (2010) considers that the integration of dispersed knowledge is far from easy or straightforward. Increasingly extensive networks impose costs in terms of coordination and learning (Howells 2008). By reducing the number of action groups, it is possible to reduce coordination and learning costs, with, for example, "pico/nano" laboratories and transitory or short-lived structures focused on very specific or short-term projects (Howells 2008). At the same time, in R&D organizations that combine basic research and systems integration, sharing agreements with the academic world can enable talents to be combined with a view to a common goal (Lam 2007).

However, in operations where disjointed disciplines need to feed off each other to achieve a common goal, it remains preferable (despite the relational power authorized by digital technology) for players to be in close spatial proximity. According to Malecki (2010), this facilitates transfers and spin-offs (Nonaka and Toyama 2005). Silicon Valley is the archetype, providing knowledge flows unavailable elsewhere (Bahrami and Evans 1995; Almeida and Kogut 1999; Saxenian 2008). But, in these attempts at creation of a teleological order, how, on the one hand, do we know where the origin of the happy idea is, since creativity results in a random process which aims at invention via a POC and, on the other hand, who is interested in the transformation of an idea, more or less formatted, into a demonstration of the latter?

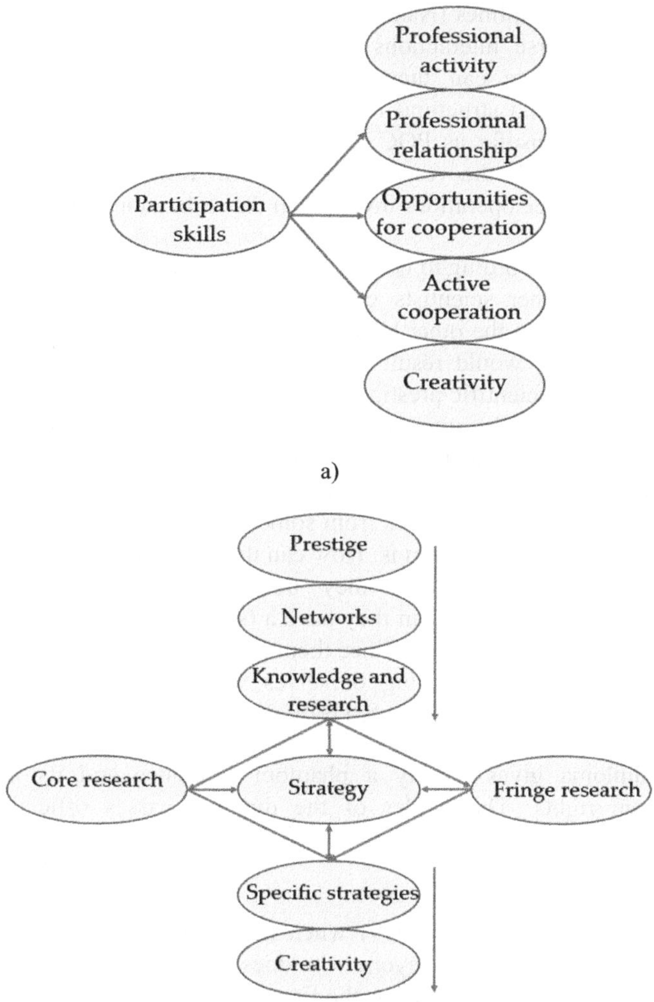

a)

b)

Figure 1.10. *Proximal creativity in cooperation (Barrett et al. 2014): (a) autonomy; (b) relationship with society*

Cooperative ventures are commonplace in scientific work. Sawyer (2006) asserts that they are a component of creativity. According to Barrett et al. (2014), some of the most important creative achievements have come from such opportunities. Indeed, many of the contacts between colleagues take place between specialists from similar but distinct disciplines (Nakamura and Csikszenmihalyi 2002), allowing for somewhat more diverse interactions and framing of thought (see Figure 1.10). Organizational resilience can therefore be linked to organizational creativity: cognitive, emotional and structural resources are important (Richtner and Löfsten 2014). Furthermore, insofar as POC requires interdisciplinary production activity, these relationships with the research team form a basic breeding ground for subsequent convergence operations involved in the realization of the POC.

On the one hand, according to Barrett et al. (2014), having multiple opportunities to cooperate with other scientists contributes to increasing the scope of their theoretical influence. On the other hand, these activities, when aimed at a link with socioeconomic circles, would result in a more modest theoretical influence. This context, in which the scientific prestige of the initiator can serve as a basic reference of trust, is likely to limit the autonomy of creative efforts through knowledge of the partners' needs. Nevertheless, the potential presence of an idea promoter in terms of POC is an important guarantee of a possible transition (see "valley of death no. 2"). And what happens if the idea comes from someone whose skills are unknown? The question that must preoccupy them is: How can they be recognized for the quality of their disruptive ideas, how can they build up support around them for entrepreneurship, in short, how can they have a (small) competent audience without having the means to recruit and convince them readily available? How can we attack this ivory tower without adorning ourselves with processes of recognition to penetrate it, conservatively, as presented by Valéry (1957):

> The diploma gives society a phantom guarantee and its holders phantom rights. The holder of the diploma passes officially for possessing knowledge, and comes to believe, in turn, that society owes him something.

Can we then trust Rivière (1919) when he writes: "Intelligence is decidedly fashionable. There is no longer anyone who does not claim its favors; no manifesto appears in which it is not advocated as the first of the virtues".

The literature on decision-making (Simon 1973) identifies three phases: intelligence, conception and choice. The decision-making process, which is really risk-taking, is highly complex and dynamic, surrounded by both uncertainty and ambiguity (Cohen 1985). Mintzberg et al. (1976) identify the following phases: identification (recognition and diagnosis routines), development of alternatives (search and design routines) and, as shown in Figure 1.11, selection (filtering, evaluation and authorization).

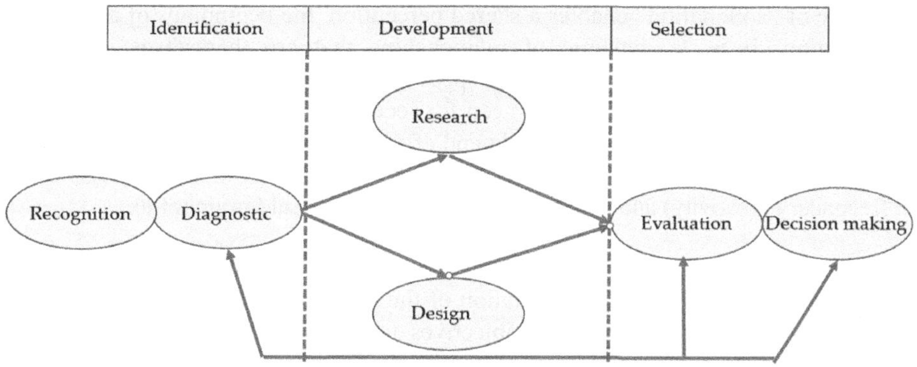

Figure 1.11. *From the initial to processing*

> We only transmit well what we are looking for, not what we already know (Duchange 2010).

> One of the difficulties, and not the least, of the evaluation of executive functions or the relative merits of executives is that there are few possibilities of directly observing the essential operations of decision-making. The fact that most executive decisions produce no direct evidence of themselves and that knowledge of these decisions can only come from the accumulation of circumstantial evidence is puzzling. They must largely be inferred from general findings in which they are only one factor, and from symptomatic indications of deviousness (Barnard 1948)[1].

> One can only agree well on misunderstandings (Valéry 1957).

Is there a right way to organize creativity? It is rather a non-issue for researchers, especially when it comes to the role of constraints, because normally, in an academic research unit, structures that are in principle creative are often associated with freedom, autonomy, a certain weakness of the rules and the scarcity of limits.

However, according to Caniëls and Rietzschel (2015), studies suggest that contraints, particularly design constraints, often stimulate creativity rather than suppress it. In the initial phase of the earliest stages of a radical innovation, the technical objectives to be considered are only present in the mind of one person, or at best a few; it's a personal "spurt". It is then that a small group allows the pre-POC (if it is feasible to set it up): generally, it belongs to the same group. This flexible

1 Author's translation from the French version of the text.

structure of "solicitation" enables a shared perception, the beginnings of a collective imagination. It is "a dynamic of relationships that are themselves defined as intellectual, emotional and affective perspectives, and actualized, i.e. put into action and into words, on the basis of the need for recognition that necessarily entails the existence of these exchanges of thought and affect" (Barus-Michel 1985). If the initial originates from a carrier, the group may subscribe to their scenario, but may also enrich it (secondary creativity) and thus become one with a proposal (group reflex).

Then, insofar as the author(s) is (are) solicited, the realization of a POC corresponds to the material manifestation of the way in which they or an associated *designer* imagine or visualize the objectives to be achieved with the design and material means to be implemented, with, as Enriquez (1992) points out, an "obsession with plenitude". It is a way of creating "a set of traces that encode meaning" (Desplet 2019). The materialization of these "technological visions" is useful for understanding the progression of the operation, especially from a technical discontinuity.

However, the desired transition is a fragile operation if it is deeply differentiated, because it depends on a large number of distinct and specialized actors, with varied and numerous requirements, which can lead to centralizing management and therefore a consequent loss of wealth of interior connections, always sensitive to the mode of operation of the unit. For the group to be productive, it is necessary to take into account the free coexistence in the cultured minds associated with the operation of the most dissimilar ideas, formations, scientific and technological cultures, the most opposed principles of knowledge, etc. Each person has tendencies and thoughts that are unaware of each other, and the goal of the action is, let us remember, to bring them together to keep their wealth, without nipping them in the bud by an overly interventionist management. Indeed, the need for regrouping is linked to the fact that "we are witnessing the disappearance of the Man who could be complete, as well as the Man who could be materially self-sufficient. Significant decrease in autonomy, depression of the feeling of mastery, corresponding increase in confidence in collaboration, etc." (Valery 1957).

That said, decision-makers who are willing to commit to radical innovations have, according to Reid et al. (2015), different objectives: benefits, efficiency, magnetism, specificity and clarity of infrastructure. "Technological vision has a significant positive impact on technical competitive advantage, rapid success with customers and ability to attract capital, as measures of rapid success".

But once the decision has been made, how do you reach the target, which is to produce a POC? This is already a design activity, which envisages the object to be created almost at the stage of the creative idea (corrected from the outset by knowledge of what is real and what is possible) or, a little later, during the realization phase, that is, at any time and at any level of the project hierarchy. This important distinction underlines the need to understand the entire design process, and consequently the creative behavior from the happy idea onwards. With ill-conditioned proposals (at least at the outset), if the objective is clear, the stages in the project's progression can vary according to project orientation (sometimes with detour), chronology (and time pressure) or hierarchy. The boundaries between these compartments are imprecise and may be entangled by numerous parameters, leading to overlaps. Stages can be based on local interests (with some thought given to the concepts that need to be implemented to fulfil the intended functions) when several people are involved in the process. In complex processes, tasks may be assigned to different disciplines.

However, the design process, which may evolve as a result of scientific and technical (or even financial) challenges, needs to be studied in its entirety – from initial analysis to detailed design. This ensures that no design sub-activity is discarded, and keeps track of changes, enabling a contrast to be drawn between stages. To carry out such an operation is to practice design to implement projects and tools dedicated to this context. This "as you go" approach also seems to echo the mechanics of creativity. Figure 1.12 illustrates the POC process.

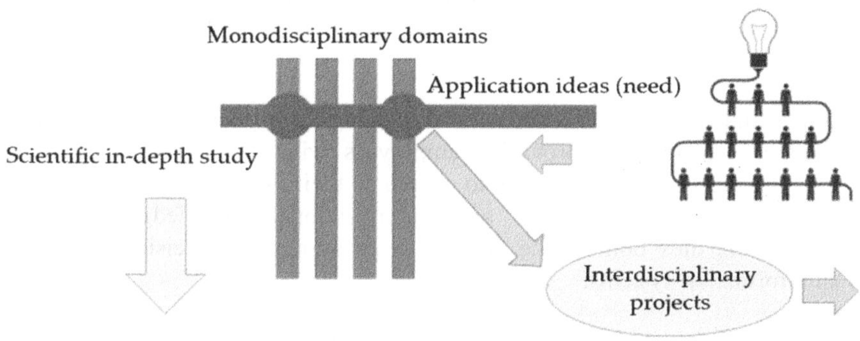

Figure 1.12. *How to create a POC*

Creative behavior, according to Guilford (1956), includes an element of divergence or convergence in which the designer creatively diverges and/or converges by exploring the principles of combinatorial solutions or their problem. To achieve their objectives, the designer can draw on their experience, and use more open-ended techniques: brainstorming, analogy-finding, identification of emergent properties, problem exploration through the co-evolution of problem spaces and solutions (Dorst and Cross 2001), following opportunistic design processes (Bender and Blessing 2004), etc.

However, the POC designer needs to know that each member of their project group is capable of carrying out tasks that contribute to the project's progress by integrating into the activities of other members with different skills (trade, scientific discipline, etc.) at the right moment. Snider et al. (2016) consider that the study of patterns, similarities, common and unusual tasks allows us to understand how people do their work, the "best practice" of expert or creative designers and the similarity or difference in the actual appearance of behaviors. In design science, formalization, networking and overall knowledge capitalization are important issues that can save time in this operation (Rias 2017).

In carrying out a POC, we come close to the principles of action research, whose aim is to produce, if possible, a methodology and a model to represent it (Baskerville 1997). They can help to enrich knowledge in the design sciences (Rias 2017). Bias is inherent in the fact that experiments are conducted by people with their own cultures and partial knowledge. Difficulties inherent in the conduct of humans can make comparisons between projects that should be based on identical protocols (Checkland and Holwell 1998). This, in turn, limits the possibility of generalizing results to other contexts. But, to quote Valéry (1957), "execution is the moment when the worker must have the greatest control over himself, because it is the best way for him to realize the impact he can have on the [social body]". If the methodology is not disseminated, it is the performance of the POC's realization work that is to be sought.

Saulais and Ermine (2020) show that knowledge is becoming a crucial resource. Knowledge is also about providing employees with the support they need to exchange knowledge, enabling them to overcome their intrinsic cognitive limits, and value and develop individual and organizational knowledge. Knowledge management is thus built at the crossroads of systems to decision systems and formal information systems.

Moreover, particularly over the past few decades, knowledge, particularly in its commercial dimension, has suddenly taken on capital, decisive importance in our society, which is described as post-industrial and largely immaterial, to constitute a new engine of growth, combined with the essential imperative of innovation. Saulais and Ermine (2020) propose, as shown in Figure 1.13, an information model called the "daisy model".

And yet, with these people committed to action, with sub-projects to explore, with a constantly redefined guideline, can we produce a robust POC? Indeed, we need to have proof of the feasibility of the idea (invention) on the market and, further still, of the company's ability to bring the innovation to fruition. This situation becomes even more complicated with startups from the academic world. "The difficulty for researcher-entrepreneurs is that, apart from scientific know-how, there is very little to demonstrate to prove the product's feasibility on the market" (Krichen and Chabaud 2019).

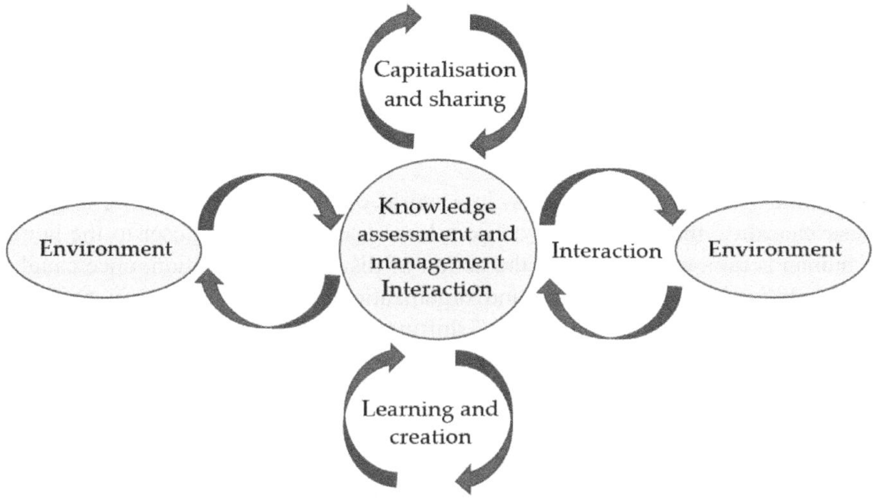

Figure 1.13. *Daisy model*

1.3. Interdisciplinarity and heuristic approach

> From the depths of the centuries, our languages echo the profound relationship between niche and appropriation, between dwelling and possession: I dwell, therefore I have (Serres 2008).

> Modern science was born when attention shifted from the search for the "what" to the investigation of the how (Arendt 1972).

> Interdisciplinarity requires translation and mediation services related to intellectual issues that go far beyond science. In my opinion, they are philosophical. They have to do with the meaning of knowledge, all knowledge, considered as the experience of the human species, aligned with its will to know and its biological survival (Hamel 2002).

Upstream of the design stage, can/should the researcher develop their responsible "quest" alone, or in concert with the "enlightened authorities" who define the framework, or with society as a whole? It is therefore easy to imagine the difficult and risky path that the application of the scientific imagination can take toward the tangible, which is already technical, and then satisfies market and socioeconomic criteria. It must move from discipline to interdisciplinarity, to integration, with aspects of attractiveness, supply, questioning for performance and thinking about possible risks, as well as recycling, of the productions it helps to bring out.

At this stage, is it not already an almost hopeless adventure, or at least one strewn with (too many) pitfalls? Particularly, if we are content with good, contextualized, peer-reviewed publications in our careers, the intermediate stage of the POC stage ensures that we do not get too far ahead of ourselves, and that we do not lose motivated scientific players along the way.

Thinking about design implies, in a world of disciplinary scientific origins and linear causality, that we need to adopt a less classical vision, closer to the humility of human action, and admit that the action of disruptive innovation, once caught up in unstabilized interdisciplinary and organizational dynamics, may in part escape the initial intention. Against this shifting backdrop, the desire for control, rationalization, standardization of behavior and the reduction of risk-taking, applied to increasingly complex systems, reinforces the risk of failure and, potentially, their difficult governance in an unsteady world by obscuring disciplinary hierarchies, tensions, questioning, the understanding of oppositions between process and content, and the qualities of interdisciplinary interactions that are so necessary to their functioning.

So, how can epistemology (the part of philosophy that studies the sciences to determine their logical origin, value and scope) be of use to the scientific and technical players involved in the POC? As this is essentially a technical adventure, the players are, for reasons of time, more users of scientific knowledge and technical tools than creators of the latter, even if within the operation new avenues of study have to be devised. But, generally speaking, the question of interfaces between disjointed areas of knowledge arises in order to achieve the goal of demonstration via a POC. These interfaces are critical, as the disciplines have different scientific issues, instruments, paradigms, languages, etc., to deal with. As Dubois and Brault (2021) and Dubois (2020) propose, for the realization of a POC, it is essential to have knowledge of the sciences and techniques that cannot be separated from its use:

> This leads both to a praxeology (philosophy of action in its methodological sense), a reflection on what scientific-technical decision-making is and to ethics (philosophy of action with reference to values: a good or just action) (Dubois 2020).

What engineering has in common with artistic activity is the creation and the improvement of artifacts. Attempting to improve technologies under time constraints and with limited resources "leads to the search for methods of maximizing success and minimizing the consumption of resources" (Dubois and Brault 2021). This is a common situation with the POC in association with the use of heuristics (the art of inventing and making discoveries by solving problems from incomplete knowledge) in fields where knowledge is partial or uncertain to bring about the best change in a situation where there are insufficient and/or unreliable scientific resources and within the limits of available resources. In this creative activity, we always need to master the basics of the disciplines involved in the project on our own or within the group, while thinking about how to establish and deal with the necessary interdisciplinarities. In this phase, the members of the project-group need to be able to deliberate or discuss the ends (what the POC is going to be used for, at least in the first instance), and not just the means (which loop around the ends).

NOTE.– A definition of interdisciplinarity:

Without proposing a definitive definition, Thompson Klein (2008) proposes seven constituent elements, which she presents in descending order:

– variability of objectives (taking into account context-dependency, valuing flexibility);

– variability of criteria and indicators (not applying criteria inherited from disciplinary thinking, but opening up to new ones, e.g. aesthetic criteria or the potential to open up new avenues of research);

– integration (importance of appropriation by all collaborators; integration of different disciplinary perspectives upstream of research);

– sociological aspects (ability to resolve conflicts, communication between participants, etc.);

– project management (effectiveness of consensus-building, analysis of task allocation, time available for meetings and discussions);

– iteration and transparency (does the program provide the means for its own evaluation by defining clear objectives and success criteria?);

– effectiveness and impact (less important, since the impact of interdisciplinary research is generally felt over the long term).

Valley of death no. 6: creating a POC within an interdisciplinary framework

1.3.1. *Theoretical considerations*

> Every body needs its members, and every art its instruments. And the moment the whole is created, so are its parts (Leonardo da Vinci 2006).

> A shared invention is produced by players from different backgrounds, with different types of knowledge brought together in shared intelligence. This production is aided by social dynamics developing the coherence of these inventions and reinventions of old or abandoned techniques (Vigezzi 2019).
>
> Translators translate data from fields other than their own into a language they're comfortable with, to make the data say things that the scientists or organizations that produced it don't want it to say. Seen another way, translators apply a model to several different materials, to produce original knowledge (Li Vigni 2021).
>
> The horizon of science remains a formidable tongue in cheek. All scientific institutions are constantly talking about interdisciplinarity. But in reality, the balance of power is merciless. Interdisciplinarity, of course, on the condition that nothing changes (Wolton 2013).

In spite of this often-burdensome context, realism must produce robust judgments or representations, at least approximate to certain phenomena that can be mastered by the disciplines (regardless of the procedures used to implement the POC). "Scientific realism is therefore a philosophical thesis that concerns the cognitive status of the products of science" (Varenne 2012). To integrate disjointed knowledge, Dalle-Nazébi (2008) points out that different concepts can be proposed for thinking about the realization of a POC, which is to bring together, in what this author calls hybrid spaces, disciplinary researchers, technicians, designers, novice individuals, etc. They are brought together to work toward a same objective (the POC project, its management, resources, etc.). These spaces take various forms, such as transaction zones (Galison 1997), trans-epistemic arenas (Knorr-Cetina 1982), boundary objects or organizations (Star and Greisemer 1989; Guston 2001; Kelly 2003; Milanovic 2006), or hybrid collectives and forums (Callon et al. 2001; Callon 2003), etc. These forms are designed to meet the need to integrate heterogeneous actors in the realistic construction of a POC project. It would even have been possible to add Bourdieu's field theory to this list, given that "the construction of the field makes it possible to establish the truth of different positions and the limits of validity of different positions (claiming truth or not) whose defenders tacitly agree, as I have indicated, to mobilize the most powerful instruments of proof or refutation that the collective achievements of their science assure them" (Bourdieu 1997).

Operations of this kind give us the opportunity to think in interdisciplinary terms, partnership and, consequently, hybridization of knowledge. The people involved in this process are responsible for developing concrete solutions, if possible directly applicable to the realization of the partially rationalized idea. For Hérard (2021), this crossover involves thinking "with" rather than "about". To thrive, interdisciplinarity needs to share the creation of a common language based on analytical skills, while enabling exchange, relationality and adaptability for project preparation through the co-production of knowledge.

But in this necessary operation, the researcher is not completely free. Hérard (2021) writes: "The researcher is also the way the institution thinks of him and the way it thinks of itself. The institution is not monolithic. It has room for maneuver. It could very well integrate the 'social form of research', accepting its productions by giving them a specific status". This important reflection reveals a context (including that of evaluation) that many researchers are reluctant to enter. Nevertheless, the *European Commission's White Paper on Communication* (EU 2006) emphasized that scientists have a duty to share their new knowledge with a wider public, and that this should be considered a priority both during projects and in subsequent impact assessment. This emphasis is on "research that matters"! So interdisciplinary work can also aim to stimulate the intellectual questioning specific to each discipline.

Li Vigni (2021) defines ""interdisciplinarity" as a social phenomenon in which different epistemic cultures (Knorr-Cetina 1999) meet and dialogue, exchanging tools, theoretical and normative frameworks, as well as know-how, and often sharing common practical goals":

> This phenomenon is therefore the manifestation of an epistemic interculturality. But it's not just a cognitive, epistemological and methodological affair. The social dimension of interdisciplinarity also includes practices, materialities, social norms, ontologies, policies and institutions. (Hackett et al. 2007)[2].

Researchers are free to choose their subject, but it is necessary to know how to approach and analyze it, and in these upstream/downstream project conditions, it is not possible to do just anything because our own activity has an effect on those of the other partners. Each member of a group works according to the theoretical presuppositions and methodology offered by each discipline (paradigms), which does not prevent them from questioning and working with other disciplines in a spirit of interdisciplinarity.

Group work is not, therefore, a place for expressing personal opinion:

> You have to start by ploughing a field, digging through archives, constructing an object of study, structuring a corpus, carrying out an initial classification of observables, then proceed to a detailed

2 Author's translation from the French version of the text.

> description of what has been collected in the field or constituted into a corpus, or even carry out experiments, then establish new classifications according to discriminating criteria (what is similar, what is different), and finally engage in interpretations. Interpretation is one of the most delicate and anxious moments for a researcher. In a constant to-and-fro between inductive and deductive reasoning, he or she must engage in internal interpretations, those made within the theoric framework of the discipline (which can lead to it being called into question), and external interpretations in terms of what other disciplines can say about the same subject (Charaudeau 2021).

An interdisciplinary project, which by its very nature involves several people and several disciplines (or even several professions), cannot be considered a sure thing; it is, in essence, a risky operation because, for various reasons, it has not been possible to pool local proposals for action.

Interdisciplinary research, cross-disciplinary thinking, collective intelligence, etc., are current trends that challenge traditional conceptual tools, which are robust and validated, but rather conservative and incremental, while drawing attention to the limiting power of any methodology. Academic discourse, which aims to expand the frontiers of intellectual thought (and this is part of its mission), seems at the same time to set its limits, a process in which an unstabilized methodology of discovery inevitably participates (is stabilization possible?). In the age of interdisciplinarity (at least in its political discourse), theory and methodology face new challenges for which creativity normally represents the keystone of a collective training process that is difficult to put in place in order to produce. It is a question of going far beyond the timid roadmap applied to additive manufacturing presented by Utela et al. (2008), by introducing trial-and-error methods, precisely because supply-side logics are no longer credible. Will we be able to find the right partners, change/open the minds of researchers, give them time, envisage acceptable funding and, above all, open and proactive decision-makers to support a so-called pre-POC operation and ensure that the idea finds its rightful place?

At this stage, specific knowledge must be combined in order to envisage new knowledge, but this knowledge is not yet usable for the POC and, of course, cannot be transposed into applications (even if we may have some ideas about the potential value of the result of this alliance between disciplines (performance, uses)). Moreover, the notion of a value network was introduced by Christensen (1997), translated as a potential for diffusion in a network. Rias (2017) recalls that the process of expansion (thanks to the reduction of the distance between technological and value potential) helps to define innovation regimes (see also Garel and Rosier (2008)). These different regimes constitute, according to Rias (2017), a typology represented in three levels in Figure 1.14.

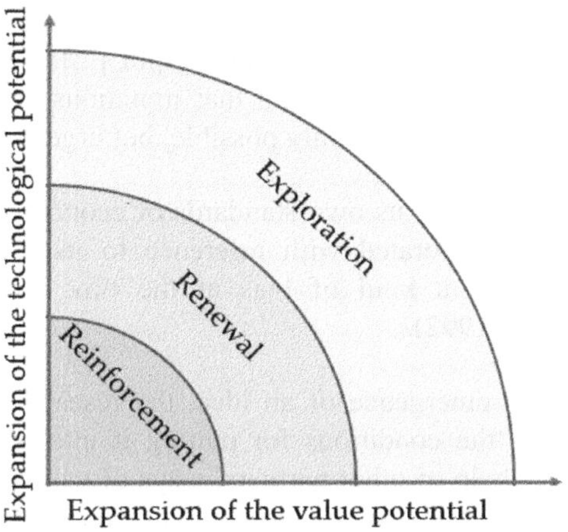

Figure 1.14. *Typology of innovation regimes*

These elements can provide food for thought and help to lend credibility to the pre-POC activity.

1.3.2. *Heuristics and interdisciplinarity*

> The interdisciplinary whole overcomes disciplinary antagonisms and the strict hierarchy of disciplines and, thanks to a circular dynamic, paves the way to a regime of associations. Although the disciplinary project distinguishes, favors and consecrates, the interdisciplinary program combines, solidarizes and demystifies. It is perhaps an advanced stage in the secularization of knowledge. The movement of knowledge implies, on a permanent basis, a displacement of frontiers, or rather the creation of a cross-border territory. [...] Until now, in the first interdisciplinary experiments, it was still the discipline that commanded the methodological movements of interdisciplinarity [...]. From now on, we need to broaden this approach by thinking in terms of the discipline within the interdisciplinary horizon. This means, once again, that interdisciplinarity is not a headlong rush, but a tireless, open-minded and mutually supportive effort to rebuild the discipline (Portella 1992).
>
> But disciplinary paradigms dominate our minds, whatever we say, and determine our mentality [...]. This vision requires not only the convergence, but also the synergy of different specialties, so that transdisciplinarity – the blurring of boundaries – retains the rigor of

> each discipline [...]. The task of our time is to create the conditions for overcoming this cultural and intellectual split [...]. Let us not forget that it is only far from equilibrium that mutations are possible, and today transformations are not only possible, but urgent (Mayor 1992).

> Old technology imposes its own standards of economic evaluation on new technology, elaborated with reference to its natural qualities, thereby introducing a kind of bias at the time of the economic calculation (Foray 1992).

To follow up on the emergence of an idea, the researcher or creator needs to refine it to characterize the conditions for putting it into practice, and to define, either alone or with the help of other partners, ways of validating and building it. If it is possible to deduce the final state without any preconceived ideas, we are talking about an incremental activity. On the other hand, if it takes a fairly large number of realizations of the phenomenon to derive an unthought-out approach, we are talking about a breakthrough. Being at the interface between disciplines, this work requires the co-construction of a shared mental representation:

> No one is capable of constructing this coherent mental representation at the interface of disciplines, nor of linking it to the various relevant theoretical corpuses, nor of putting it to the test of facts using appropriate methods. So how can we build it together? (Bourdin 2022).

Scientists consider it "self-evident that disciplinary hybridization is the key to scientific and technical innovation. Today, everyone knows that we never innovate, except on the margins of established disciplines. Indiscipline is even invoked as the condition for real discoveries and there is no shortage of examples of serendipity, which testify to the fecundity of chance and the mavericks who know how to seize it" (Besnier and Perriault 2013).

But then, the complexity generally associated with interdisciplinarity is a situation with a variety of different elements, forms and natures interacting dynamically.

Innovation moves the complexity of a situation by artificially modifying a relationship between interacting elements (it imposes an original path between elements in the form of an imposed convergence). When it is disruptive, with its characteristics of interaction between multiple players, uncertainty, unpredictability, co-evolution of the project and its context, etc., it then becomes part of a problem of complexity. Some believe that it requires an evolving strategy to emerge (invent as you go).

It is this vision (André 2017c) that we wish to question based on disruptive innovations involving the convergence of disjoint disciplines (see Figure 1.15).

The heuristic approach can be defined as an empirical, approximative, intuitive, but rapid and daring/risky method for finding a possible solution to a question for which not all knowledge is stabilized, and thus gaining access to a field of expertise. The result is a partially deductive field of expertise, perhaps accelerating the resolution of the problem for which one (of several) possible solutions is highlighted. No idea is born "good": it merely represents a new hypothesis, which may, perhaps, become so.

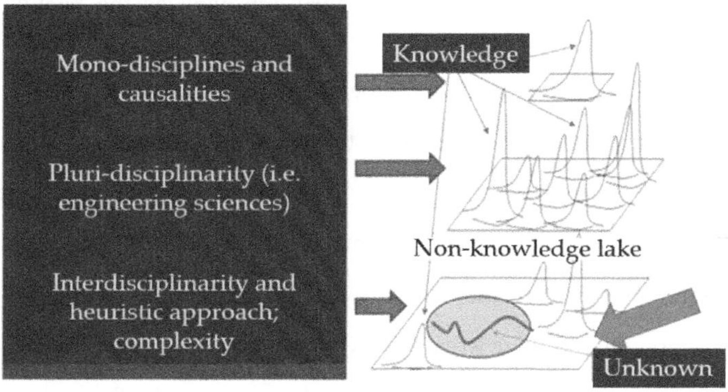

Figure 1.15. *Fields of expertise (the approximate Gaussian lines represent the scientific achievements of previous research work; depending on the form of expertise, the field is covered (causal, algorithmic) or partially covered (heuristic))*

Zhou et al. (2022) attempt to envisage more or less quantitative methods to remove epistemic uncertainties (insufficient information), with the aim of making the best decision possible (Wen et al. 2017). These authors report on numerous methods, ranging from fuzzy logic (Zeng et al. 2013), interval theory (Wang et al. 2018), evidence theory (Wang et al. 2022) and so on. These models pose a few problems, however: the need for data to be made at least partially quantitative (with the risk of induction being absent?) and the assumption of a unimodal relationship with a single optimum.

NOTE.– Interdisciplinary thesis:

According to Calenda (2021), an interdisciplinary doctoral dissertation calls for the mobilization of a well-established approach to literature analysis, but also (and above all) to answer the following questions:

– "How does the decision to embark on a doctoral thesis under the auspices of interdisciplinarity work upstream?

– How is the choice of thesis topic made with regard to interdisciplinarity?

– How is the fog of departure managed?

– How do you construct a provisional question and 'transform' it into a clear problem?

– How do you develop the working hypotheses that underpin the interdisciplinary challenge?

– How do you choose an interdisciplinary approach as the backdrop for your work?

– Which data collection tools should be used?

– What analyses should be undertaken to clarify the results of the thesis?

– What are the twists and turns when writing an interdisciplinary thesis?

– How do you design an interdisciplinary research system?

– How do you write an introduction to your presentation?

– How do you defend your work wisely?

– How can you enhance the value of your work from a post-doctoral perspective?"

The last point is very important, as most of the experts who evaluate an "interdisciplinary" publication have a less broad cultural background, which often raises questions.

1.3.3. *Some constraints to overcome*

> It is absurd for human beings [...] to hope that maybe another Newton might some day arise, to make intelligible to us in terms of natural laws unordered by any intention, even the genesis of a blade of grass [...] (Kant 2007).

> According to chemist George Cowan, [...], the institute was to persuade "certain people to become generalists, educate people to broaden their gaze across several disciplines", to become "Renaissance Men of the 21st century", able to deal with the chaotic and inelegant real world, which science doesn't really address (Li Vigni 2021).

> I don't usually delve into the literature of other fields of research [...] that's a mortifying approach, because then you get stuck in the details of the question and you become indoctrinated by the spirit of that other field and by the things its representatives think are possible or impossible (Li Vigni 2021).

> Individualism establishes as its first principle that individuals are called upon to develop their faculties to the full extent to which they are susceptible; that these faculties should be limited only to the extent necessary for the maintenance of tranquility and public safety; and that no one is obliged, with regard to his opinions, beliefs and doctrines, to submit to an intellectual authority outside himself (Constant 2002).

Exogenous factors can have a major impact on the endogenous innovation process by accelerating or delaying the emergence of a profitable technological breakthrough. These factors can have several origins – historical, technical, local political (research unit, region, state) and financial – and influence risk-taking.

Salgian et al. (2013) remind us that for many citizens of the western world, despite acceptable trust, science and technology have an image problem: "Scientists and engineers are generally perceived as 'intellectuals' who spend their lives doing things no one else understands. When asked to list creative activities, people rarely mention science or engineering and instead suggest the arts and humanities". Negative views on the teaching of "hard" sciences by undergraduates (difficult and boring subjects) need not be considered, and it is only after several years that the fields taught can be seen as supporting their creativity (Qin 2009). At present, however, the majority of interdisciplinary endeavors are the result of adventures based on individual encounters, affinities, opportunities and chance. It is all about people (Coutellec 2009).

This aspect of inter-individual relations is particularly critical and may be, independently of financial support, responsible for many failures linked to the lack of knowledge of people likely to help define or participate in the epistemology of POC. The intellectual construction of the latter requires the presence and proactive action of a mediator, who acts as an interface between representatives of the disciplines involved in the project, enabling a field of possibilities to be explored. Their role also enables them to integrate and pool considerations from the various disciplines for action (participation in an interactive approach that constantly reinterprets the sciences and techniques associated with the project, which is systemic by nature). No mediator, no salvation!

What is more, we have noted that involvement in a risk-taking operation that corresponds to fostering inventive creativity could be seen as unappealing to disciplinary researchers. This is all the more true when the latter are formatted to produce publications, admittedly of a good scientific level, but confined to the deepening of their restricted field. This leads to an inverse vision of the world, known as the "laboratization" of the world (Callon et al. 2001; Callon 2003). Figure 1.16 illustrates this context. It goes without saying that recruiting this type of personality is not normally one of the actors sought for inter-disciplinary work (although...). So, while the contribution of interdisciplinary results can be seen as a strategic element of project preparation (Gaubinger and Rabl 2014), it is generally essential to save as much time as possible and, naturally, for the lowest cost (Reid and de Brentani 2004), which can constitute another limiting factor of the operation.

In addition, according to Marseille News (2021), interdisciplinary research is more difficult to fund, which limits the commitment to these risky forms critical research. The article reminds us of this:

> Scientists argue that the constant struggle for funding undermines good work by encouraging researchers to over-promise and engage in questionable practices, incentivizing publication in the best journals, discouraging replication of existing work and stifling creativity and intellectual risk-taking.

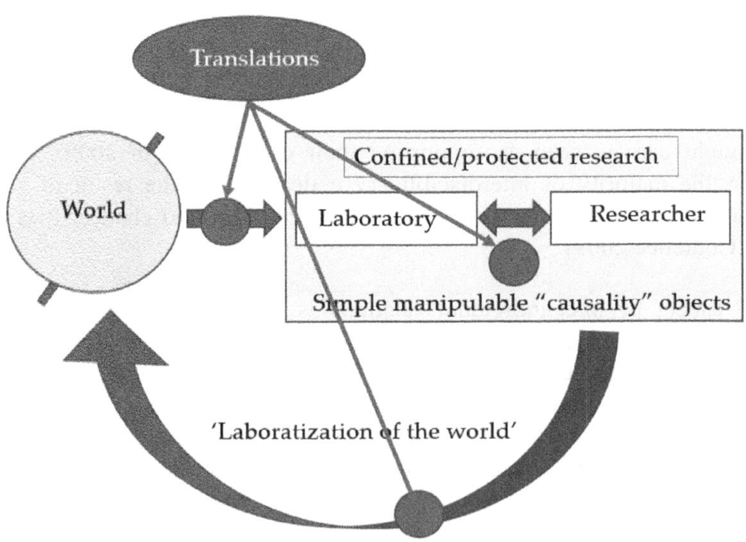

Figure 1.16. *Narrowing associated with radical translations*

Thus, while governments, through their plans and major research programs, make financial commitments to "political narratives" that focus on the major trends that pose problems (and require serious responses) for society (e.g. global warming and its consequences), researchers can strive for "scientific excellence" in a way that is not necessarily measured by real-world impact (Walter et al. 2007; Sedlacko et al. 2013; Barré 2015; Rau et al. 2018).

But, even if these plans are recent, Oakey (2015) teaches us that post-1945 economic theory (Arrow and Debreu 1954; Debreu 1959) generally considered that economic systems worked best when supply and demand were in equilibrium, "implying that this stability is a state of optimal efficiency towards which sectoral and general markets should evolve and return when they have been disturbed". This position is out of step with the Schumpeterian approach, which exploits the principle of "creative destruction" (Schumpeter 1939, 1942), which needs new ideas to unfold. He is followed in this by Christensen et al. (2015). The demand for invention can then be held back by the holders of the equilibrium, the status quo, who try to prevent competition (formation of oligopolies or monopolies).

Risk-taking, which must be supported by the hierarchy, supposes that the fact of not anticipating the common project should be one of the essential springs of collective creativity. Indeed, Piaget (1967) wrote: "It is the most radical way of annihilating all discourse to isolate each reality from all attachment to the rest; for it is through the mutual combination of forms that discourse is born to us". New ideas, based on partial knowledge of the real and the possible, should be born and take shape along a path that is yet to be constructed. Going through phases of disorder, accepting detours, not knowing, not judging, not excluding, leaving room for chance and freeing the imagination: it is in these conditions – sometimes playful, sometimes insecure – that divergent thinking and creativity must find expression. Thanks to a wide range of methodologies, made up of work processes and situations, collective intelligence, combined with the project, should make it possible to fertilize real interdisciplinary innovation.

1.3.4. *In practice*

Valley of death no. 7: supporting your way to the POC and beyond

> We shouldn't worry about whether what we're doing is science, but whether what we're doing is likely to touch on a truth. You don't have to choose between art and science. You choose art in the sense of doing what you have to do well in order to try to show what you think is a truth, rather than knowing whether you're responding to modes of estimation that mean that what you're doing will be recognized as science. We mustn't start from a system of opposition, nor from the idea that art (or literature) is on one side and science on the other (Rancière 1994).

> If creation is individual, the preparatory work for creation is group-based. To bring his work to fruition, the creator needs the support of his group. The group is a creative matrix. Without it, the individual cannot fully deploy his potential (Chouvier 2017).

> From different horizons comes the idea that order, disorder and organization must be thought of together. The mission of science is no longer to banish disorder from its theories, but to deal with it. It is no longer to dissolve the idea of organization, but to conceive and introduce it to federate fragmented disciplines (Romian 2000).

At this stage, when we have a technical file, we will have to examine usage scenarios further. A clarification of the idea, as for the POC, is based on a multidisciplinary and integrative approach built by people other than the researcher-designers. The scenarios must be extended at least to their organizational aspect. Multiple iterations with potential users (or their representatives) are necessary "during the stage of clarifying the problem, identifying expectations, developing intermediate artefacts and developing the various prototypes to be tested in real-life situations. These multiple iterations are costly in terms of time, human and financial resources" (Pascal and Rouby 2017).

The environment is critical because you have to find the right skills and manage interdisciplinary cooperation, with the risk that the project develops its own language, its own practices and its own frame of reference and has difficulty serving the innovation. In addition, the participants must include in their activity the formulation of a strategy that requires an understanding of the evolution of the competitive environment (competing technologies) and a vision of the strategic positioning to be established for the construction of a competitive advantage. However, how can we be sure that after having worked on shared and emerging representations, we are following the right strategy when there are many indeterminacies, when the robust definition of the field of possibilities is not possible? Who will take the related risks?

So, once we have drawn up the functional specifications for the POC and beyond, in principle we have a document that can be organized according to the classic principles of "who could do what, how, when" (see GANT, for example). At this stage, however, a number of pitfalls can arise:

– have the people and resources to achieve the objective;

– have the financing in time;

– avoid misappropriation of ideas by unscrupulous evaluators;

– the difficulty of good industrial protection.

Krugman wrote (see Daniel (2021)): “Productivity isn’t everything, but in the long term it’s almost everything. A country’s ability to improve its standard of living over the long term depends almost entirely on its ability to increase output per worker”. This is indeed the basis on which support is defined, and the project sponsors are going to need it! These remarks raise the question of the financial “size” of the POC. If the supervisory authorities do not seem to be very interested in the project, should the POC or its simple project be presented to industrialists likely to be interested in the new Holy Grail, with confidentiality clauses which, in the event of a dispute, would force the matter to go to court? A link with a French company is all well and good, but it is even more risky when you go international! So, to support researchers in France, SATTs were invented (technology transfer acceleration companies). At the heart of this creative act lies the transfer process. It is more or less mastered by the subject and, according to Almudever (2017), the transfer operates links between registers of activities and experiences:

> By displacement (of knowledge, attitudes, know-how, values, gestures, etc.) from one domain to another, from one period of socialization to another; by the impression of a representation or meaning on other representations and meanings; by the projection of affects, the circulation of emotions from one situation to another.

The assumption made (and sometimes challenged) is that the promoter of an idea is facing a competent person.

Selecting ideas from the POC is then an important question of evaluation based on different methods and tools (Carrier and Gélinas 2011), ranging from a simple multi-criteria evaluation grid to sophisticated computer programming tools, or even lighter attitudes. Yet it is essential for the owner of the POC and/or idea, particularly for the teams responsible for handling the aftermath of the operation (Kornish and Hutchison-Krupat 2017). Due to a lack of culture or a lack of motivation to explore a subject too modestly, there is a risk of “missing out” on an idea that could have high added value (Teglborg et al. 2013). The author, in another context (André 2016), has experienced such a situation. But when vagueness reigns, we can expect support for “circumstantial” operations with financial commitments to projects that bring only modest added value to the company.

However, nepotism is not at work and there is no objective reason to follow Lepage’s (2015) opinion:

> Subsidies are preferably directed towards sectors where the problems are politically the most apparent, where there are organized pressure groups with significant experience and effective weapons to interest public authorities in their problems and impose their conception of what to do to solve these problems.

NOTE.– SATT:

According to the French Ministry of Research (MESRI 2017b), the 13 Technology Transfer Acceleration Companies (SATT) act as a bridge between research laboratories and companies, financing the project maturation and POC phases. These one-stop shops for technology transfer enhance the efficiency and the competitiveness of the innovation system. By January 1, 2022, they had detected and analyzed 16,357 innovative projects, protected 3602 priority patents and granted 1452 licenses to companies. They also participated in the creation of 672 startups (Conecticus 2022). SATT Sayens (2020), which covers the Grand Est and Bourgogne-Franche-Comté regions (12.5% of the French population), is said to have been responsible for the creation of 22 startups, including 12 from programs, 335 registered titles (not defined in the report) and 346 "portfolio" projects. It is possible that locally (the author is concerned by this SATT), the number of startups, which would be 3.3% of the total, raises questions relative to the size of the population concerned. Furthermore, according to the French Senate's (2017) report, not all SATTs necessarily meet the criteria that are nevertheless the conditions for their success, with procedures that sometimes remain perceived as complex and cumbersome. For the time being, the SATTs' track record remains mixed in terms of the financial "windfall" made available for technology transfer.

The technological innovation envisaged by the POC results from a coupling between the dynamics of exploring innovative concepts (the idea) and the dynamics of knowledge building, which can be complex and interconnected. Expertise professions should be called upon to transform themselves to support intensive innovation strategies (Le Masson et al. 2006). "They can only support these strategies at the cost of learning made more demanding and difficult by current organizational conditions" (Charue-Duboc et al. 2020). The risk, then, is of focusing on the short term and solving problems of the same type as those encountered elsewhere (and therefore known to the expert evaluator). The latter can then make a series of proposals, one after the other, with little time to go into the issues in depth. Figure 1.17 shows the progression from the POC (or, in some cases, the pre-POC).

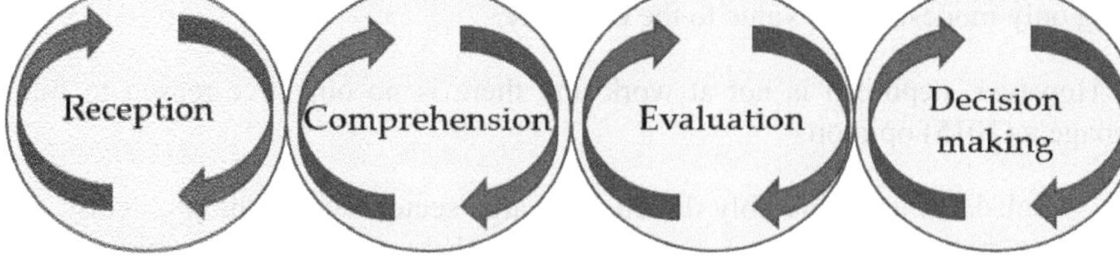

Figure 1.17. *From pre-POC (or POC) to maturation*

NOTE.– CNRS prematuration:

The aim of this program is to provide financial support and guidance for emerging projects with high innovation, technological or usage potential. "Starting by observing the basic principle of an innovative idea, we develop the concept of the technology and/or its application and provide an experimental demonstration" (CNRS 2021). In particular, the work supported within this framework enables us to validate a POC. It can therefore be used upstream of maturation by a SATT by providing financial resources to achieve POC. In contrast to the previous situation, proposals are evaluated by a committee made up of representatives from CNRS and the socioeconomic world, who assess the dossiers on the basis of innovation and commercialization criteria. It interviews project leaders and provides recommendations for the next stages of the project.

Apart from these two situations, with which the author is quite familiar, there are numerous state organizations (Carnot institutes or competitveness clusters; see Battini (2015) for example), a 2022 call for proposals to accelerate technology transfer skills and professions of the future: adapting the training system for young people and employees to the professions of tomorrow" (SGPI 2022), technology transfer centers (CRITT), etc. Financial support can also be provided by the regions (MESRI 2017a) – to the tune of a billion euros in 2017 – universities, engineering schools, towns and cities, etc. In fact, money is undoubtedly available, but with a high number of windows (and filters)! We probably need to be a professional when it comes to requesting (as is often the case for calls for proposals in H2020 or Horizon Europe). But, in the favorable hypothesis that we have been funded and we end up with industrial protection, the ordeal, as we will understand, is not over yet.

NOTE.– National Academy of Inventors (NAI):

Today, researchers who make inventions and successfully manage their commercialization are celebrated for the impact their work has had on both their fields of study and the economy. The NAI has a website[3] which lists its members, with links to their biographies and information on the selection process.

1.4. Conclusion

> To succeed in discovery, we need to have unlikely approaches. If we think like everyone else and bring people together to discover something who have the same way of thinking, the chance of discovering something is very low [...]. Discovery is as beautiful as

3 Available at: https://academyofinventors.org/.

> the chance meeting on a dissection table of a sewing machine and an umbrella, in other words, something improbable (Lautréamont 2011).

> For every complex problem, there is an answer that is clear, simple and wrong (Mencken, quoted in Sturmberg and Topolsky (2014)).

> For us to be certain, we must necessarily add to the imagination something which is reasoning (Spinoza 1954).

> The subject must be able to reconnect with their original thaumaturgic power in order to overcome the prohibitive obstacles that paralyze them. Many people prevent themselves from creating on the pretext that they are incompetent or even useless, when in fact this inoperability is rooted in self-deprecation. By regaining self-confidence, the subject is able to overcome their creative inhibitions (Chouvier 2017).

There is no single type of creative person committed to disruption. But if they have followed the "path of the cross" to innovation, it is because they plan to be a motivated player in efficiency looking for support and, if possible, drawing on the maximum available resources, at least to get to POC and industrial protection. These creatives are often considered to be located "between two worlds". Gaglio (2011) refers to them as marginals-secants or passers, with a "dual cultural belonging".

What has been shown is that they have the ability to create and maintain networks, and to bring partners on board with them, motivating them in a truly trusting environment, with the will to build collectively. Creativity gives rise to other forms of creativity, such as moving toward reality by redefining tasks to adjust to concrete situations within a shared framework, innovating by singling out activities, "the mobilization and detour of means aimed at overcoming obstacles encountered" (Gaudart 2017).

In the general representation of the project associated only with the idea, many unknowns are possible: then, with the limited means of the group associated with it, it will be necessary to convince hierarchies, financiers, principals, etc., which imposes a robust mastery of the "resistance of reality". The partial indeterminism of activity therefore opens up to the unforeseen (seen as an obstacle, failure, accident, disavowal); but as a chance to open up to creation, to inventiveness. This work, as defined in this chapter, is a real adventure. It is true that, seen from a distance from the field, a form of predictability is needed, which identifies areas that become priorities, forgetting the need for subsidiary knowledge that contributes to the development of these areas. In what has been presented, the creativity is rather spontaneous, not claiming at least at the beginning, real traceability. Paris (2017) writes:

> Creation implies the integration of inputs, many of them, producing outputs, without any link being established between the latter and the former. To put it another way, the recognition of the abundance necessary for all creation means that organizations renounce all forms of traceability, and therefore all linear vision or quest for optimization.

But, after this spark, everything is structured to go further, otherwise the idea dies out.

Perhaps to conclude this chapter, we should return to these few important sentences:

> To cut short the "danse macabre des principes" that ends up paralyzing teaching and research institutions, will we be able to encourage new undisciplined people to oppose the conformism of academia with style and freedom? Will we be able to tolerate that, at the end of a fruitful research career, we can hear one of these undisciplined people serenely express that "the most important thing for him was not to get bored" (Dupuy 2013), or to preserve "the elasticity, inventiveness and fantasy that make the company of youth so delightful"? (Feyerabend 1976)

From a more societal point of view, "it can't be stressed enough that the key here is to make the French breeding ground more business-friendly, rather than inventing 'industrial policies'. The fundamental task is to eliminate the well-known handicaps: the overgrowth of our public sphere, the piling up of regulations, the precautionary principle and taxes on production and capital" (Fontanet 2022). It is up to entrepreneurs to take risks and develop a framework of high confidence by looking for opportunities that others do not see. But this risk-taking requires stable (yet not oppressive!) social foundations.

1.5. References

Aggeri, F., Teulier, R., Duru, M. (2009). Des modèles théoriques à la rencontre des pratiques à travers les instruments. In *Concevoir et construire la décision. Démarches en agriculture, agroalimentaire et espace rural*, de Turckheim, E., Hubert, B., Messéan, A. (eds). Quæ, Paris.

Aghion, P., Antonin, C., Bunel, S. (2020). *Le pouvoir de la destruction créatrice*. Odile Jacob, Paris.

Allison, P.D. and Long, J.S. (1990). Departmental effects of scientific productivity. *American Sociological Review*, 55, 469–478.

Almeida, P. and Kogut, B. (1999). Localization of knowledge and the mobility of engineers in regional networks. *Management Science*, 45, 905–917.

Almudever, B. (2017). Créativité, création et personnalisation au travail. In *La créativité au travail*, Amado, G., Bouilloud, J.P., Lhuilier, D., Ulmann, A.L. (eds). Erès, Paris.

Alter, N. (2001). *L'innovation ordinaire*. PUF, Paris.

Amabile, T.M. (1996). *Creativity in Context*. Westview Press, Boulder.

Amabile, T.M., Conti, R., Coon, H., Lazenby, J., Herron, M. (1996). Assessing the work environment for creativity. *Academy of Management Journal*, 39, 1154–1184.

André, J.C. (2013). Towards a socially responsible research (SRR) charter in engineering sciences at CNRS level. *International Journal of Technoethics*, 4, 39–51.

André J.C. (2016). Désordres dans la troisième dimension ; un cheminement entre formation et interdisciplinarité. *Histoire de la recherche contemporaine*, 5, 62–72.

André, J.C. (2017a). *From Additive Manufacturing to 3D/4D Printing 1: From Concepts to Achievements*. ISTE Ltd, London, and John Wiley & Sons, New York.

André, J.C. (2017b). *From Additive Manufacturing to 3D/4D Printing 2: Current Techniques, Improvements and their Limitations*. ISTE Ltd, London, and John Wiley & Sons, New York.

André, J.C. (2017c). *From Additive Manufacturing to 3D/4D Printing 3: Breakthrough Innovations: Programmable Material, 4D Printing and Bio-printing*. ISTE Ltd, London, and John Wiley & Sons, New York.

André, J.C. (2020a). L'innovation de rupture, élément d'une éthique du futur applicable à la santé environnementale ? *Environnement, risques et santé*, 19, 404–416.

André, J.C. (2020b). Engagement solidaire et déconfinement ? *Environnement, risques et santé*, 19, 238–239 [Online]. Available at: https://www.jle.com/fr/covid19-engagement-solidaire-deconfinement.

Arendt, H. (1972). *La crise de la culture*. Folio, Paris.

Arrow, K. and Debreu, G. (1954). Existence of an equilibrium for a competitive economy. *Econometrica*, 22, 54–71.

Arthur, W.B. (1996). Increasing returns and the new world of business. *Harvard Business Review*, 74, 100–109.

Aubin, J.P. (2013). Indisciplinés de toutes les disciplines, dispersez-vous ! In *Le gouvernement des technosciences ; gouverner le progrès et ses dégâts depuis 1945*, Pestre, D. (ed.). La Découverte, Paris.

Bahrami, H. and Evans, S. (1995). Flexible re-cycling and high-technology entrepreneurship. *California Management Review*, 37, 62–89.

Baillargeon, N. (2012). Comment "penser" critique ? *Afcas Magazine* [Online]. Available at: https://www.acfas.ca/publications/magazine/2012/11/comment-penser-critique.

Barnard, C. (ed.) (1948). *Organization and Management: Selected Papers*. Harvard University Press, Boston.

Barré, R. (2015). Développement durable et système français de recherche : vraies difficultés ou faux-semblants ? In *Un demi-siècle d'environnement entre science, politique et prospective. En l'honneur de Jacques Theys*, Barré, R., Lavoux, T., Piveteau, V. (eds). Quæ, Paris.

Barrett, J.D., Vessey, W.B., Griffith, J.A., Mracek, D., Mumford, M. (2014). Predicting scientific creativity: The role of adversity, collaborations, and work strategies. *Creativity Research Journal*, 26, 39–52.

Barrier, C. (2011). La science en projets : financements sur projet, autonomie professionnelle et transformations du travail des chercheurs académiques. *Sociologie du Travail*, 53, 515–536.

Barus-Michel, J. (1987). *Le sujet social*. Dunod, Paris.

Baskerville, R.L. (1997). Distinguishing action research from participative case studies. *Journal of Systems and Information Technology*, 1, 24–43.

Bassereau, J.F., Charvet-Pello, R., Faucheu, J., Delafosse, D. (2015). Les objets intermédiaires de conception/design, instruments d'une recherche par le design. *Sciences du Design*, 2, 48–63.

Battini, P. (2015). La problématique de l'innovation en France. *Vie & sciences de l'entreprise*, 200, 119–141.

Bauman, Z. (2006). *La vie liquide*. Le Rouergue, Arles.

Bender, B. and Blessing, L. (2004). On the superiority of opportunistic design strategies during early embodiment design. *Proceedings of DESIGN 2004: The 8th International Design Conference*, Dubrovnik.

Berry, C. (1981). The Nobel scientists and the origins of scientific achievement. *British Journal of Sociology*, 32, 381–391.

Besemer, S.P. and O'Quin, K. (1999). Confirming the three-factor creative product analysis matrix model in an American sample. *Creativity Research Journal*, 12, 287–296.

Besnier, J.M. (ed.) (2013). Seul le désordre est créateur. Pour en finir avec les bataillons disciplinaires. *Revue Hermès*, 67(3), 25–31.

Besnier, J.M. and Perriault, J. (2013). Introduction générale. *Revue Hermès*, 67(3), 13–15.

Blumenberg, H. (1993). *La légitimité des temps modernes*. Gallimard, Paris.

Boardman, C. and Bozeman, B. (2006). Implementing a "bottom-up" multi-sector research collaboration: The case of Texas air quality collaboration. *Economics of Innovation and New Technology*, 15, 51–69.

Bonasso, S.G. (2007). Inquiry, discovery, invention, and innovation. The personal experience of technology generation and transfer in engineering and scientific research. *Leadership and Management in Engineering*, 7, 141–150.

Bonnardel, N. and Marmèche, E. (2005). Towards supporting evocation processes in creative design: A cognitive approach. *International Journal of Human-Computer Studies*, 63, 35–442.

Bouiss, O. (2020). Qu'est-ce qu'une "crise" ? *The Conversation* [Online]. Available at: https://theconversation.com/quest-ce-quune-crise-136026#:~:text=La%20crise%20est%20un%20%C3%A9veil,conscience%20soudaine%20de%20la%20complexit%C3%A9.

Boulekbache-Mazouz, H. (2014). Introduction. In *Recherches en design : processus de conception, écriture et représentations*, Leleu-Merviel, S. and Boulekbache-Mazouz, H. (eds). ISTE Editions, London.

Bourdieu, P. (ed.) (1997). Pour une sociologie clinique du champ scientifique. In *Les usages sociaux de la science*. Quæ, Versailles.

Bourdin, S. (2022). Quelle acceptabilité sociale des technologies disruptives ? *The Conversation* [Online]. Available at: https://theconversation.com/quelle-acceptabilite-sociale-des-technologies-disruptives-173337.

Bourque, C.J. (2016). La perception de l'innovation chez les chercheurs canadiens des sciences dites dures. *Quaderni*, 90, 81–95.

Bousset, L. (2012). Travail d'ébauche interdisciplinaire : peut-on parler simplement d'un processus complexe ? *Natures Sciences Sociétés*, 20, 448–454.

Briens, F. (2015). La Décroissance au prisme de la modélisation prospective. Exploration macroéconomique d'une alternative paradigmatique. PhD Thesis, ENSMP, Paris.

Brisset, N. (2014). Performer par le dispositif ? Un retour critique sur la théorie de la performativité. *L'Année sociologique*, 64, 217–246.

Brown, D.C. (2012). Creativity, surprise & design: An introduction and investigation. *ICDC2012: 2nd International Conference on Design Creativity*, 75–86, The Design Society, Glasgow.

Bunge, M. (1997). Mechanism and explanation. *Philosophy of the Social Sciences*, 27, 410–465.

Bunge, M. (2007). Did Weber practice the objectivity he preached? In *Max Weber's "Objectivity" Reconsidered*, McFalls L. (ed.). University of Toronto Press, Toronto.

Calenda (2021). Faire une thèse interdisciplinaire : les ficelles et les enjeux [Online]. Available at: https://calenda.org/946287.

Callon, M. (2003). Laboratoires, réseaux et collectifs de recherche. In *Encyclopédie de l'innovation*, Mustar, P. and Penan, H. (eds). Economica, Paris.

Callon, M. and Ferrary, M. (2006). Les réseaux à l'aune de la théorie de l'acteur-réseau. *Sociologies pratiques*, 13, 37–44.

Callon, M., Lascoumes, P., Barthe, Y. (2001). *Agir dans un monde incertain, essai sur la démocratie technique*. Le Seuil, Paris.

Caniëls, M.C.J. and Rietzschel, E.F. (2015). Organizing creativity: Creativity and innovation under constraints. *Creativity and Innovation Management*, 24, 184–196.

Carlile, P.R. (2004). Transferring, translating and transforming knowledge: An integrative framework for managing knowledge across boundaries. *Organization Science*, 15, 555–568.

Carrier, C. and Gélinas, S. (2011). *Créativité et gestion. Les idées au service de l'innovation*. Presses de l'université du Québec, Montreal.

Castoriadis, C. (1975). *L'institution imaginaire de la société*. Le Seuil, Paris.

Charaudeau, P. (2021). Polémique sur l'"islamo-gauchisme" : revenir à l'esprit de la recherche. *The Conversation* [Online]. Available at: https://theconversation.com/polemique-sur-l-islamo-gauchisme-revenir-a-lesprit-de-la-recherche-155889.

Charue-Duboc, F., Gastaldi, L., Bertin, E. (2020). Des communautés d'experts internes comme facilitateur de l'innovation. *Revue française de gestion*, 287, 81–98.

Checkland, P. and Holwell, S. (1998). Action research: Its nature and validity. *Systemic Practice and Action Research*, 11, 9–21.

Chesbrough, H.W. (2003). *Open Innovation: The New Imperative for Creating and Profiting from Technology*. Harvard Business School Press, Boston.

Chesbrough, H.W. and Crowther, A.K. (2006). Beyond high tech: Early adopters of open innovation in other industries. *R&D Management*, 36, 229–236.

Chouvier, B. (2017). Groupalité et processus sublimatoire. In *La créativité au travail*, Amado, G., Bouilloud, J.P., Lhuilier, D., Ulmann, A.L. (eds). Erès, Paris.

Christensen, C.L. (1997). *The Innovator's Dilemma*. Harvard Business School Press, Boston.

Christensen, C.M., Raynor, M.E., McDonald, R. (2015). What is disruptive innovation? *Harvard Business Review*, 93, 44–53.

Clark, R.A. and Rice, G.A. (1982). Family constellations and eminence: The birth orders of Nobel Prize winners. *Journal of Psychology*, 110, 281–287.

CNRS (2021). Programme Pré-maturation du CNRS [Online]. Available at: https://www.inp.cnrs.fr/sites/institut_inp/files/download-file/Cadrage%20et%20crit%C3%A8res%20de%20s%C3%A9lection.pdf.

Cohen, I.B. (1985). *Revolution in Science*. Belknap Press, Boston.

Conecticus (2022). Le Réseau SATT publie de nouveaux résultats de performance [Online]. Available at: https://www.conectus.fr/en/node/579.

Constant, B. (2002). *Le libéralisme inquiet*. Michalon, Paris.

Coutellec, L. (2009). Biologies face à la modélisation et à l'interdisciplinarité. *Natures Sciences Sociétés*, 17, 73–75.

Cowen, T. (2011). *The Great Stagnation: How American are All the Low-Hanging Fruit of Modern History, Got Sick and Will (Eventually) Feel Better*. Dutton, New York.

Cramarégeas, F. (2021). L'interdisciplinarité dans l'enseignement secondaire en France (1970-2018) : une légitimation contrariée dans les discours, les configurations institutionnelles et les pratiques enseignantes. PhD Thesis, Université de Bordeaux [Online]. Available at: https://tel.archives-ouvertes.fr/tel-03483282/document.

Csikszentmihalyi, M. (2014). Implications of a systems perspective for the study of creativity. In *Handbook of Creativity*, Sternberg, R.J. (ed.). Cambridge University Press, Cambridge.

Csikszentmihalyi, M. and Csikszentmihalyi, I.S. (eds) (2006). *A Life Worth Living: Contributions to Positive Psychology*. Oxford University Press, Oxford.

Cummings, J.N. and Kiesler, S. (2005). Collaborative research across disciplinary and organizational boundaries. *Social Studies of Science*, 35, 703–722.

D'Aveni, R.A. (1994). *Hypercompetition*. Free Press, New York.

Dalle-Nazébi, S. (2008). Objet et acteurs de recherche. La montée en expertise de locuteurs de langues des signes. *Revue d'anthropologie des connaissances*, 2, 63–96.

De Gaulejac, V. (2012). *La recherche malade du management*. Quæ, Versailles.

De Vinci, L. (2006). 15 avril 1452/2 mai 1519 : Carnets. In *Inventions*, Anna, H. (ed.). Parragon Books Ltd, Bath.

Debreu, G. (1959). *Theory of Value: An Axiomatic Analysis of Economic Equilibrium*. Yale University Press, New Haven.

Demoly, F. and André, J.C. (2022). *4D Printing 1: Between Disruptive Research and Industrial Applications*. ISTE Ltd, London, and John Wiley & Sons, New York.

Despret, V. (2019). *Habiter en oiseau*. Actes Sud, Arles.

Disis, M.L. and Slattery, J.T. (2010). The road we must take: Multidisciplinary team science. *Science Translational Medicine*, 2(22), 22cm9.

Doran, J. and Ryan, G. (2017). The role of stimulating employees' creativity and idea generation in encouraging innovation behavior in Irish firms. *Irish Journal of Management*, 36, 32–48.

Dorst, K. and Cross, N. (2001). Creativity in the design process: Co-evolution of problem-solution. *Design Studies*, 22, 425–437.

Doz, Y., Santos, J., Williamson, P. (2001). *From Global to Meta-national: How Companies Win in the Global Economy*. Harvard Business School Press, Boston.

Dubois, M.J.F. (2020). *L'humain en devenir : au commencement était la technique*. ISTE Editions, London.

Dubois, M.J.F. and Brault, N. (2021). *Manuel d'épistémologie pour l'ingénieur*. Editions Matériologiques, Paris.

Ducasse, I. (2011). *Les Chants de Maldoror*. Penguin Classics, London.

Dunbar, K. (1995). How scientists really reason: Scientific reasoning in real-world laboratories. In *The Nature of Insight*, Sternberg, R.J. and Davidson, J.E. (eds). MIT Editions, Boston.

Dunbar, K. (2000). How scientists think in the real world: Implications for science education. *Journal of Applied Developmental Psychology*, 21, 49–58.

Dupuy, J.P. (2013). Le capitalisme a besoin de se croire immortel pour exister. *Revue de la régulation* [Online]. Available at: http://journals.openedition.org/regulation/10048.

Eckert, C.M., Wyatt, D.F., Clarkson, P.J. (2009). The elusive act of synthesis: Creativity in the conceptual design of complex engineering products. *Proceedings of the Seventh ACM Conference on Creativity and Cognition*, Berkeley.

Eliade, M. (1963). *Aspects du mythe.* Gallimard, Paris.

Encyclopédie Universalis (1995). *Article "Innovation".* Encyclopédie Universalis, Paris.

Enriquez, E. (1992). *L'organisation en analyse.* PUF, Paris.

European Union (2006). Livre blanc sur la communication de la Commission européenne [Online]. Available at: https://ec.europa.eu/councils/bx20060615/white_paper_fr.pdf.

Fei, Y.M. and Lin, C.J. (2013). Digital knowledge searching depth and individual creativity. *Computer Applications in Engineering Education*, 21, 230–236.

Feyerabend, P. (1976). On the critique of scientific reason. In *Method and Appraisal in the Physical Sciences: The Critical Background to Modern Science, 1800–1905*, Howson, C. (ed.). Cambridge University Press, Cambridge.

Floricel, S. and Miller, R. (2003). An exploratory comparison of the management of innovation in the new and old economies. *R&D Management*, 33, 501–525.

Fontanet, X. (2022). Relocalisation : comment font les autres pays ? *Contrepoints* [Online]. Available at: https://www.contrepoints.org/2022/08/16/374265-comment-reussir-la-relocalisation.

Foray, D. (1992). Choix de la technique, rendements croissants et processus historiques. In *La technoscience. Les fractures du discours*, Pradès, J. (ed.). L'Harmattan, Paris.

Ford, C.M. (1996). A theory of individual creative action in multiple social domains. *Academy of Management Review*, 21, 1112–1142.

Ford, C.M. (2000). Creative developments in creativity theory. *Academy of Management Review*, 25, 284–289.

France Angels (2022). Livre blanc. Les Business-Angels, élément moteur de l'économie française [Online]. Available at: https://mcusercontent.com/48981545f014cb67c77b7ba1e/files/915a3efb-673d-58f4-7afb-4a996e7dbc92/Livre_Blanc_2022_FA.pdf.

French Senate (2017). Les SATT : des structures de valorisation de la recherche publique qui doivent encore faire la preuve de leur concept [Online]. Available at: http://www.senat.fr/rap/r16-683/r16-683_mono.html#toc0.

Gaglio, G. (ed.) (2011). Les formes contemporaines de l'innovation. In *Sociologie de l'innovation*. PUF, Paris.

Galison, P. (ed.) (1997). The trading zone: Coordinating action and belief. In *Image and Logic: Instruments and Arguments in Twentieth Century Physics*. Chicago University Press, Chicago.

Garel, G. and Rosier, R. (2008). Régimes d'innovation et exploration. *Revue française de gestion*, 187, 127–144.

Gaubinger, K. and Rabl, M. (eds) (2014). Structuring the front end of innovation. In *Management of the Fuzzy Front End of Innovation*, Gassmann, O. and Schweitzer, F. (eds). Springer International Publishing, New York.

Gaudart, C. (2017). La créativité, une appropriation du temps. In *La créativité au travail*, Amado, G., Bouilloud, J.P., Lhuilier, D., Ulmann, A.L. (eds). Erès, Paris.

Gero, J.S. (1998). Conceptual design as a sequence of situated acts. In *Artificial Intelligence in Structural Engineering*, Smith, I. (ed.). Springer, Berlin.

Geslin, P. (2017). *Inside Anthropotechnology: User and Culture Centered Experience*. ISTE Ltd, London, and John Wiley & Sons, New York.

Gibson, C.B. and Birkinshaw, J. (2004). The antecedents – Consequences and mediating role of organizational ambidexterity. *Academy of Management Journal*, 47, 209–226.

Gille, B. (1978). *Histoire des techniques*. Gallimard, Paris.

Gingras, Y. (2014). *Les dérives de l'évaluation de la recherche. Du bon usage de la bibliométrie*. Raisons d'agir, Paris.

Giust-Desprairies, F. (2003). *L'imaginaire collectif*. Erès, Paris.

Global Europe Anticipation Bulletin (2022). Évaluation de nos anticipations pour 2021 : 77% de succès [Online]. Available at: https://geab.eu/evaluation-de-nos-anticipations-pour-2021-77-de-succes/?utm_source=sendinblue&utm_campaign=GEAB_-_La_lettre_danticipation_du_3_fvrier&utm_medium=email.

Godefridi, T. (2021). Les 3 moteurs du capitalisme, les 3 principes de la liberté. *Contrepoints* [Online]. Available at: https://www.contrepoints.org/2021/12/31/256086-3-moteurs-capitalisme?mc_cid=4e7d6f9e52&mc_eid=1b4a9d2e39.

Gonzales, L.D. (2014). Framing faculty agency inside striving universities: An application of Bourdieu's theory of practice. *Journal of Higher Education*, 85 193–218.

Goux-Baudiment, F. (2002). Une nouvelle étape de la prospective : la prospective opérationnelle. PhD Thesis, Université pontificale grégorienne de Rome, Rome.

Gramaccia, G. (2008). *Démocratie participative et communication territoriale. Vers la microreprésentativité*. L'Harmattan, Paris.

Guilford, J.P. (1950). Creativity. *American Psychologist*, 5, 444–454.

Guilford, J.P. (1956). The structure of intellect. *Psychological Bulletin*, 53, 267–293.

Guilford, J.P. (1959). Traits of creativity. In *Creativity and its Cultivation*, Anderson, H.H. (ed.). Harper Collins, New York.

Guilford, J.P. (1967). Creativity: Yesterday, today and tomorrow. *The Journal of Creative Behavior*, 1, 3–14.

Guillot, A. and Meyer, J.A. (2014). *Poulpe-Fiction. Quand l'animal inspire l'innovation*. Dunod, Paris.

Gusdorf, G. (1968). Interdisciplinarité (connaissance). *Encyclopaedia Universalis*, vol. 8, Paris.

Guston, D.H. (2001). Boundary organizations in environmental policy and science: An introduction. *Science, Technology & Human Values*, 26, 399–408.

Guzzetti, L. (1998). Euroscientia forum. *European Commission*, 1, 85–86.

Hackett, E.J., Amsterdamska, O., Lynch, M., Wajcman, J. (eds) (2007). *The New Handbook of Science and Technology Studies*. The MIT Press, Boston.

Hamel, J. (2002). La pédagogie comme pivot de l'interdisciplinarité. *Revue internationale d'éducation de Sèvres*, 30, 143–151.

Harroche, A. (2021). Gouverner par les inégalités : la mise en œuvre d'une initiative d'excellence dans l'enseignement supérieur et la recherche. PhD Thesis, Institut d'études politiques de Sciences Po, Paris.

Harvey, J.F., Naggar, R., Cohendet, P., Simon, L. (2013). Gérer les idées pour mieux innover. *Gestion*, 38, 25–34.

Hatchuel, A. (2006). *Quelle analytique de la conception ? Parure et pointe en design. Le design essais sur des théories et des pratiques*. Editions du regard, Paris.

Hatchuel, A. and Weil, B. (1992). *L'expert et le système*. Economica, Paris.

von Hayek, F. (1952). *The Counter-Revolution in Science. Studies on the Abuse of Reason*. The Free Press, New York.

Heidegger, M. (ed.) (1958). Que veut dire penser ? In *Essais et conférences*. Gallimard, Paris.

Heitz, A. (2022). Albert Einstein et les brevets. *Contrepoints* [Online]. Available at: https://www.contrepoints.org/2022/03/14/423406-albert-einstein-et-les-brevets.

Helson, R. and Pals, J.L. (2000). Creative potential, creative achievement, and personal growth. *Journal of Personality*, 68, 1–27.

Henry, C. (1998). Le "je" intellectuel et le "jeu" interdisciplinaire. *Le Genre humain*, 33, 155–170.

Hérard, M. (2021). Synthèse des séminaires "Recherches actions, recherches collaboratives, recherches participatives". *Maison des Sciences de l'Homme Paris Nord* [Online]. Available at: https://hal.archives-ouvertes.fr/hal-03406012/document.

Hirsch, J.E. (2005). An index to quantify an individual's scientific research output. *Proceedings of the National Academy of Science*, 102, 16569–16572.

Hohberger, J. (2016). Diffusion of science-based inventions. *Technological Forecasting and Social Change*, 104, 66–77.

Holahan, P.J., Sullivan, Z.Z., Markham, S.K. (2014). Product development as core competence: How formal product development practices differ for radical, more innovative, and incremental product innovations. *Journal of Product Innovation Management*, 31, 329–345.

Howells, J. (2008). New directions in R&D: Current and prospective challenges. *R&D Management*, 38, 241–252.

Hubert, M., Chateauraynaud, F., Fourniau, J.M. (2012). Les chercheurs et la programmation de la recherche : du discours stratégique à la construction de sens. La reconfiguration des universités selon l'OCDE. Economie du savoir et politique de l'innovation. *Actes de la recherche en sciences sociales*, 148, 68–73.

Ininou, Y. and Loilier, T. (2021). La sélection des idées incrémentales et radicales. Les enseignements d'une étude de cas. *Innovations*, 66, 139–175.

Jansen, R.A., Raferty, A.N., Grifths, T.L. (2021). A rational model of the Dunning–Kruger effect supports insensitivity to evidence in low performers. *Nature Human Behavior*, 5, 756–763.

Journal du Net (2021). Brevet d'invention : fonction, intérêt, démarches, coût [Online]. Available at: https://www.journaldunet.fr/management/guide-du-management/1200257-le-brevet-d-invention/.

Jung, C.G. (1964). *The Development of Personality*. Routledge, London.

Jünger, F.G. (2018). *La perfection de la technique*. Allia, Paris.

Jurdant, B. (2011). La vulgarisation scientifique, quel(s) effet(s) pour le chercheur ? Journées d'études. PhD Thesis, Université de Dijon, Dijon [Online]. Available at: https://ocim.fr/2011/01/la-vulgarisation-scientifique-quels-effets-pour-le-chercheur/.

Kant, I. (2007). *Critique of Judgement*. Oxford University Press, Oxford.

Kelly, S.E. (2003). Public bioethics and publics: Consensus, boundaries, and participation in biomedical sciences policies. *Science, Technology & Human Values*, 28, 339–364.

Keynes, J.M. (2021). *La vertu et le bonheur : suivi de possibilités économiques pour nos petits-enfants*. Les Belles Lettres, Paris.

Kijkuit, B. and van den Ende, J. (2007). The organizational life of an idea: Integrating social network, creativity and decision-making perspectives. *Journal of Management Studies*, 44, 863–882.

Kin, V., Rollet, A., Senkel, M.P., Jan, F. (2021). Génération et sélection d'idées dans le processus d'innovation : cas de la méthodologie Delphi dans la logistique 4.0. *Innovations*, 66, 109–138.

King, A. (1999). *Géopolitique du chaos*. Folio, Paris.

Klein, J.T. (2010). *Creating Interdisciplinary Campus Cultures: A Model for Strength and Sustainability*. Jossey-Bass, San Francisco.

Knorr-Cetina, K. (1982). Scientific communities or transepistemic arenas of research? A critique of quasi-economic models of science. *Social Studies of Science*, 12, 101–130.

Knorr-Cetina, K. (1999). *Epistemic Cultures: How the Sciences Make Knowledge*. Harvard University Press, Boston.

Koestler, A. (1964). *Le cri d'Archimède*. Calmann-Lévy, Paris.

Kornish, L.J. and Hutchison-Krupat, J. (2017). Research on idea generation and selection: Implications for management of technology. *Production and Operations Management*, 26, 633–651.

Koro-Ljungberg, M. (2002). Constructions of high academic achievement through the analysis of critical events. *Gifted Child Quarterly*, 46, 209–233.

Krichen, K. and Chabaud, D. (2019). Une analyse comparée et critique des modèles des spin-offs universitaires. *Marché et organisations*, 34, 61–86.

Kruger, J. and Dunning, D. (1999). Unskilled and unaware of it: How difficulties in recognizing one's own incompetence lead to inflated self-assessments. *Journal of Personality and Social Psychology*, 77, 1121–1134.

Krugman, P. (2021). Jean-Marc Daniel : l'entrepreneur-innovateur. *Sociétal* [Online]. Available at: https://www.societal.fr/jean-marc-daniel-lentrepreneur-innovateur.

Lam, A. (2007). Knowledge networks and careers: Academic scientists in industry–university links. *Journal of Management Studies*, 44, 993–1016.

Larousserie, D. (2022). Le président du CNRS affiche une vision pitoyable de la motivation des chercheurs [Online]. Available at: https://blogs.mediapart.fr/francois-brechignac/blog/230222/le-president-du-cnrs-affiche-une-vision-pitoyable-de-la-motivation-des-cher-cheurs.

Le Chatelier, H. (1925). *Science et industrie*. Flammarion, Paris.

Le Masson, P., Weil, B., Hatchuel, A. (2006). *Les processus d'innovation. Conception innovante et croissance des entreprises*. Hermès-Lavoisier, Paris.

Le Moigne, J.L. (2013). Sur l'intelligence de la complexité, en entrelaçant épistémologie et pragmatique. In *Intelligence de la complexité. Epistémologie et pragmatique*, Le Moigne, J.L. and Morin, E. (eds). Hermann, Paris.

Leleu-Merviel, S. and Boulekbache-Mazouz, H. (eds) (2014). *Recherches en design : processus de conception, écriture et représentations*. ISTE Editions, London.

Lepage, H. (2015). *Libéralisme et liberté d'expression*. Texquis, Paris.

Leroi-Gourhan, A. (1973). *Milieu et techniques*. Albin Michel, Paris.

Les Actualités (2021). Ce que promet l'"économie des créateurs" et ce qu'elle fait réellement [Online]. Available at: https://lesactualites.news/nouvelles/ce-que-promet-l-economie-des-createurs-et-ce-quelle-fait-reellement/.

Lévy, A. (1997). *Sciences cliniques et organisations sociales*. PUF, Paris.

Lewin, K. (1951). Field theory in social sciences. In *Selected Theoretical Papers*, Cartwright, D. (ed.). Harper & Row, New York.

Li Vigni, F. (2021). Cinq types de travail scientifique “interdisciplinaire”. *Natures Sciences Sociétés*, 29, 130–140.

Link, A.N. and Scott, J.T. (2020). Distance between scientific knowledge and applications. *Diffusion of Science Based Invention Economics of Innovation and New Technology*, 29, 489–500.

Lubart, T.I. (1999). Creativity across cultures. In *Handbook of Creativity*, Sternberg, R.J. (ed.). Cambridge University Press, Cambridge.

Lubart, T.I. (2001). Models of the creative process: Past, present and future. *Creativity Research Journal*, 13, 295–308.

Maffesoli, M. (2000). *L'instant éternel*. La Table Ronde, Paris.

Magni, M., Proserpio, L., Hoegl, M., Provera, B. (2009). The role of team behavioral integration and cohesion in shaping individual improvisation. *Research Policy*, 38, 1044–1053.

Magnusson, P.R., Wästlund, E., Netz, J. (2016). Exploring users' appropriateness as a proxy for experts when screening new product/service ideas. *Journal of Product Innovation Management*, 33, 4–18.

Maher, M.L. and Poon, J. (1996). Modelling design exploration as co-evolution. *Computer Aided Civil Infrastructure Engineering*, 11, 195–209.

Maher, M.L. and de Silva Garza, A.G. (2006). *Co-evolutionary Design of Structural Layouts: A Potentially Creative Solution?* University of Sydney Press, Sydney.

Malecki, E.J. (2010). Global knowledge and creativity: New challenges for firms and regions. *Regional Studies*, 44, 1033–1052.

Mandard, M. (2018). Les usages du brevet : vers une définition du brevet comme mécanisme de gouvernance des activités d'innovation des organisations. *Innovations*, 57, 165–187.

Marseille News (2021). Une nouvelle approche du financement de la recherche scientifique peut-elle débloquer l'innovation ? [Online]. Available at: https://www.marseillenews.net/une-nouvelle-approche-du-financement-de-la-recherche-scientifique-peut-elle-debloquer-linnovation.html.

Maunoury, J.L. (1968). *La genèse des innovations : la création technique dans l'activité de la firme*. PUF, Paris.

Mayor, F. (ed.) (1992). Préface. In *Entre savoirs – l'interdisciplinarité en acte : enjeux, obstacles, résultats*. Erès, Paris.

McDougall, J. (1978). *Plaidoyer pour une certaine anormalité*. Gallimard, Paris.

Mehrotra, P., Arora, A., Krishnamurthy, S. (2021). 3 common fallacies about creativity. *Harvard Business Review* [Online]. Available at: https://hbr.org/2021/11/3-common-fallacies-about-creativity.

Mencken, H.L. (2014). For every complex problem, there is an answer that is clear, simple and wrong: And other aphorisms about medical statistical fallacies. *Journal of Evaluation in Clinical Practice* 20, 1017–1025.

Menger, P.M. (2001). Les temps, les causes et les raisons de l'action. In *Le modèle et le récit*, Grenier, J.Y., Grignon, C., Menger, P.M. (eds). Éditions de la Maison des Sciences de l'Homme, Paris.

MESRI (2017a). Les sociétés d'accélération du transfert de technologies (SATT) [Online]. Available at: https://www.enseignementsup-recherche.gouv.fr/fr/les-societes-d-acceleration-du-transfert-de-technologies-satt-47688.

MESRI (2017b). Le financement de la R&T par les Collectivités Territoriales [Online]. Available at: https://publication.enseignementsup-recherche.gouv.fr/eesr/FR/EESR12_R_47/le_financement_de_la_r_t_par_les_collectivites_territoriales/.

Milanovic, F. (2006). Travail d'articulation et organisations-frontières dans la recherche urbaine française. *Social Science Information/Information sur les sciences sociales*, 45, 109–138.

Miller, H. (1957). *Ils étaient vivants et ils m'ont parlé*. Gallimard, Paris.

Minel, S. (2003). Démarche de conception collaborative et propositions d'outils de transfert de données métier. PhD Thesis, ENSAM, Paris.

Mintzberg, H., Raisinghani, D., Theoret, A. (1976). The structure of "unstructured" decision processes. *Readings in Decision Support System*, 133–160 [Online]. Available at: http://www.sietmanagement.fr/wp-content/uploads/2016/04/Mintzberg1976.pdf.

de Montaigne, M. (2020). *Dictionnaire amoureux de Montaigne*. Plon, Paris.

Moscovici, S. (1976). *La psychanalyse, son image, son public*. PUF, Paris.

Mumford, M.D. (2000). Managing creative people: Strategies and tactics for innovation. *Human Resource Management Review*, 10, 313–351.

Mumford, M.D. and Gustafson, S.B. (1988). Creativity syndrome: Integration, application, and innovation. *Psychological Bulletin*, 103, 27–43.

Mumford, M.D., Whetzel, D.L., Reiter-Palmon, R. (1997). Thinking creatively at work: Organization influences on creative problem solving. *Journal of Creative Behavior*, 31, 7–17.

Mumford, M.D., Marks, M.A., Connelly, M.S., Zaccaro, S.J., Reiter-Palmon, R. (2000). Development of leadership skills: Experience and timing. *Leadership Quarterly*, 11, 87–114.

Mumford, M.D., Scott, G.M., Gaddis, B. (2003). Leadership in scientific organizations. In *Scientific Research Effectiveness: The Organizational Dimension*, Hurley, R.J. (ed.). Kluwer, Dordrecht.

Mumford, M.D., Connelly, S.M., Scott, G., Espejo, J., Sohl, L.M., Bedell, K.E. (2005). Career experiences and scientific performance: A study of social, physical, life, and health sciences. *Creativity Research Journal*, 17, 105–129.

Musso, P., Ponthou, L., Seulliet, S. (2007). *Fabriquer le futur 2. L'imaginaire au service de l'innovation*. Village mondial, Paris.

NAC (2014). Convergence: Facilitating transdisciplinary integration of life sciences, physical sciences, engineering, and beyond [Online]. Available at: http://nap.edu/18722.

Nakamura, J. and Csikszentmihalyi, M. (2002). Catalytic creativity: The case of Linus Pauling. *American Psychologist*, 56, 337–341.

National Center for Science and Engineering Statistics (NCSES) (2021). Understanding unsuccessful innovation. Working Paper, NCSES 21-201 [Online]. Available at: https://www.nsf.gov/statistics/2021/ncses21201/ncses21201.pdf.

Naudon, F. (2013). Comment le profane joue en faveur du décloisonnement. In *Le gouvernement des technosciences ; gouverner le progrès et ses dégâts depuis 1945*, Pestre, D. (ed.). La Découverte, Paris.

Nietzsche, F. (2013). *Beyond Good and Evil*, translated by Helen Zimmern [Online]. Available at: https://www.gutenberg.org/files/4363/4363-h/4363-h.htm.

Nonaka, I. and Toyama, R. (2005). The theory of the knowledge-creating firm: Subjectivity, objectivity and synthesis. *Industrial and Corporate Change*, 14, 419–436.

Nooteboom, B., Van Haverbeke, W., Duysters, G., Gilsing, V., van den Oord, A. (2007). Optimal cognitive distance and absorptive capacity. *Research Policy*, 36, 1016–1034.

Oakey, R. (2015). Schumpeterian economics: Some observations on the relevance of his theoretical contributions to the management of industrial research and development. *R&D Management*, 45, 386–396.

Ollivier, B. (2001). Enjeux de l'interdisciplinarité. *L'année sociologique*, 51, 337–354.

Paris, T. (2017). La dimension laborieuse de la création. De la modernité d'un archaïsme. In *La créativité au travail*, Amado, G., Bouilloud, J.P., Lhuilier, D., Ulmann, A.L. (eds). Erès, Paris.

Parmentier, G. (2020). Comment la créativité peut aider à surmonter la crise du Covid-19. *The Conversation* [Online]. Available at: https://theconversation.com/comment-la-creativite-peut-aider-a-surmonter-la-crise-du-covid-19-134859.

Parmentier, G., Szostak, B.L., Rüling, C.C. (2018). Introduction au dossier thématique. Créativité organisationnelle : quels enjeux en management stratégique dans un contexte mondialisé ? *Management International*, 22, 12–18.

Parmentier, G., Paris, T., Gandia, R. (2021). De la créativité à l'innovation : paradoxes, articulation et enjeux. *Innovations*, 66, 5–16.

Pascal, B. (2000). *Pensées*. Le livre de poche, Paris.

Pascal, A. and Rouby, E. (2017). Les scénarios d'usage comme support aux méthodologies de Recherche en Design Science dans le cas d'invention. *Systèmes d'information & management*, 22, 47–78.

Pasquinelli, E. (2022). Enfants : l'esprit critique, une qualité innée, à aiguiser dès le plus jeune âge. *The Conversation* [Online]. Available at: https://theconversation.com/enfants-lesprit-critique-une-qualite-innee-a-aiguiser-des-le-plus-jeune-age-132714.

Perrot, M.D., Rist, G., Sabelli, F. (1992). *La mythologie programmée : l'économie des croyances dans la société moderne*. PUF, Paris.

Pestre, D. (ed.) (2014). Du gouvernement du progrès technique et de ses effets. In *Le gouvernement des technosciences ; gouverner le progrès et ses dégâts depuis 1945*. La Découverte, Paris.

Piaget, J. (1967). *Logique et connaissance scientifique*. Gallimard, Paris.

Pigasse, M. (2014). *Éloge de l'anormalité*. Plon, Paris.

Pinheiro Neves, J. (2011). Pour comprendre les nouvelles liaisons digitales : le concept d'individuation chez Carl Jung et Gilbert Simondon. *Sociétés*, 111, 105–114.

Portella, E. (1992). Introduction. In *Entre savoirs – L'interdisciplinarité en acte : enjeux, obstacles, résultats*. Erès, Paris.

Poulantzas, N. (1976). *Political Power and Social Classes*. Humanity Press, Atlantic Highlands.

Przyluski, J. (1942). *L'évolution humaine*. PUF, Paris.

Qin, B. (2009). Computer science as creativity. *The Big Red Data Blog* [Online]. Available at: http://bigreddata.blogspot.com/2009/08/computer-science-ascreativity.html.

Quan, H., Chen, X., Lan, Y., Luo, X., Kubiak, R., Bonnet, N., Paux, G. (2020). Applications of Bayesian analysis to proof-of-concept trial planning and decision making. *Pharmaceutical Statistics: The Journal of the Pharmaceutical Industry*, 19, 468–481.

Rancière, J. (1994). Histoire des mots, mots de l'histoire. *Communications. L'écriture des sciences de l'Homme*, 58, 87–101.

Rau, H., Goggins, G., Fahy, F. (2018). From invisibility to impact: Recognizing the scientific and societal relevance of interdisciplinary sustainability research. *Research Policy*, 47, 266–276.

Ray, J. (1967). *Cinq histoires intégrales*. Gérard Editions, Verviers.

Raynaud, D. (2021). *Sociologie fondamentale. Étude d'épistémologie*. Éditions Matériologiques, Paris.

Reid, S. and De Brentani, U. (2004). The fuzzy front end of new product development for discontinuous innovations: A theoretical model. *Journal of Product Innovation Management*, 21, 170–184.

Reid, S., Roberts, D., Moore, K. (2015). Technology vision for radical innovation and its impact on early success. *The Journal of Product Innovation Management*, 32, 593–609.

Revault d'Allonnes, M. (2012). *La crise sans fin. Essai sur l'expérience moderne du temps.* Le Seuil, Paris.

Rhodes, M. (1961). An analysis of creativity. *Phi Delta Kappan*, 42, 305–310.

Rias, A.L. (2017). Créativité par et pour la fabrication additive : proposition d'une méthodologie outillée. PhD Thesis, ENSAM, Paris.

Richtner, A. and Löfsten, H. (2014). Managing in turbulence: How the capacity for resilience influences creativity. *R&D Management*, 44, 137–151.

Rivière, J. (1919). Le parti de l'intelligence. *La Nouvelle Revue Française*, 13, 612–618.

Romian, H. (2000). *Pour une culture commune : de la maternelle à l'université. Des problématiques complexes. Transformer l'école et la culture scolaire.* Hachette Education, Paris.

Root-Bernstein, R.S. (2003). The role of enculturation in scientific innovation: A case study of combinational chemistry at Parke-Davis. In *Scientific Research Effectiveness: The Organizational Dimension*, Hurley, J. (ed.). Kluwer, Dordrecht.

Root-Bernstein, R.S., Bernstein, M., Garnier, H. (1995). Correlations between avocations, scientific style, work habits, and professional impact of scientists. *Creativity Research Journal*, 8, 115–137.

Roqueplo, P. (1983). *Penser la technique ; pour une démocratie concrète.* Le Seuil, Paris.

Rousseau, L. (1996). La méthode Delphi : une liaison heureuse entre la pratique et la recherche en gestion. *Revue Organisation*, 5, 27–47.

Runco, M.A. (2003). Commentary on personal and potentially ambiguous creativity: You can't understand the butterfly unless you (also) watch the caterpillar. *Creativity Research Journal*, 15, 137–142.

Saint Sernin, B. (1979). *Le décideur.* Gallimard, Paris.

Salgian, A., Nakra, T.M., Ault, C., Wang, Y. (2013). Conducting robots – Bridging the gap between science, technology and the arts in the undergraduate curriculum. In *IEEE Integrated STEM Education Conference (ISEC)*, Princeton.

Sandstrom, U., van den Besselaar, P.A.A., Dorta-González, P. (2016). Quantity and/or quality? The importance of publishing many papers. *PLoS One*, 11, e0166149.

SATT Sayens (2020). Rapport d'activité 2019 [Online]. Available at: https://www.yumpu.com/fr/document/read/63664765/rapport-dactivite-sayens-2019.

Saulais, P. and Ermine, J.L. (2020). Le management des connaissances inventives : du patrimoine intellectuel inventif à l'innovation. *Marché et organisations*, 39, 59–90.

Sawyer, R.K. (2006). Educating for innovation. *Thinking Skills and Creativity*, 1, 41–48.

Saxenian, A. (2006). *The New Argonauts: Regional Advantage in a Global Economy*. Harvard University Press, Boston.

Schumpeter, J. (1939). *Business Cycles: A Theoretical, Historical and Statistical Analysis of the Capitalist Process*. McGraw-Hill, New York.

Schumpeter, J. (1942). *Capitalism, Socialism and Democracy*. Harper & Row, New York.

Sedlacko, M., Reisch, L., Scholl, G. (2013). Sustainable food consumption: When evidence-based policy making meets policy-minded research. Introduction to the special issue. *Sustainability: Science, Practice Policy*, 9, 1–6.

Serres, M. (2008). *Le mal propre. Polluer pour s'approprier*. Le Pommier, Paris.

Serres, M. (2013). Non, le réel n'est pas découpé en créneaux.... In *Le gouvernement des technosciences ; gouverner le progrès et ses dégâts depuis 1945*, Pestre, D. (ed.). La Découverte, Paris.

Sfez, L. (2002). *Technique et idéologie*. Le Seuil, Paris.

SGPI (2022). Trois dispositifs emblématiques lancés pour accélérer le transfert de technologies, les stratégies de développement des établissements d'enseignement supérieur et de recherche et la formation aux métiers d'avenir [Online]. Available at: https://www.gouvernement.fr/trois-dispositifs-emblematiques-lances-pour-accelerer-le-transfert-de-technologies-les-strategies-de.

Shai, O., Reich, Y., Hatchuel, A., Subrahmanian, E. (2013). Creativity and scientific discovery with infused design and its analysis with C–K theory. *Research in Engineering Design*, 24 201–214.

Sigaut, F. (1988). Haudricourt et la technologie. In *La technologie, science humaine. Recherches d'histoire et d'ethnologie des techniques*, Haudricourt, A.G. (ed.). Editions de la Maison des Sciences de l'Homme, Paris.

Silberstein, M. (ed.) (2013). Conclusion. Science(s) et matérialisme(s) : examen des conditions d'une synonymie. In *Matériaux philosophiques et scientifiques pour un matérialisme contemporain : sciences, ontologie, épistémologie*. Éditions Matériologiques, Paris.

Silberzahn, P. (2022). Le marché est une institution vivante : il faut réhabiliter Adam Smith. *Contrepoints* [Online]. Available at: https://www.contrepoints.org/2022/02/22/422092-le-marche-est-une-institution-vivante-il-faut-rehabiliter-adam-smith.

Simon, H.A. (1973). The structure of ill structured problems. *Artificial Intelligence*, 4, 181–201.

Simondon, G. (1989). *L'individuation psychique et collective*. Aubier, Paris.

Smith, A. (2000). *Recherche sur la nature et les causes de la richesse des nations*. Economica, Paris.

Snider, C., Dekoninck, E., Culley, S. (2016). Beyond the concept: Characterizations of laterstage creative behavior in design. *Research in Engineering Design*, 27, 265–289.

Sosa, R. and Gero, J.S. (2003). Design and change: A model of situated creativity in IJCAI creativity workshop [Online]. Available at: http://papers.cumincad.org/data/works/att/bdfd.content.09884.pdf.

Spencer, H. (2010). *The Right to Ignore the State*. Freedom Press, London [Online]. Available at: https://www.gutenberg.org/files/34649/34649-pdf.pdf.

Spencer, H. (2017). *First Principles*. John Childs and Son, London [Online]. Available at: https://www.gutenberg.org/cache/epub/55046/pg55046-images.html.

Sriskandarajah, N., Christnesen, D., Kobayashi, S., Sønksen, H. (2006). Building effective learning networks among researchers and stakeholders in the Danish context. In *Changing European Farming Systems for a Better Future: New Visions for Rural Areas*, Langeveld, H. and Roling, N. (eds). Academic Publishers, Wageningen.

Star, S.L. and Greisemer, J.R. (1989). Institutional ecology, "translations" and boundary objects: Amateurs and professionals in Berkeley's Museum of Vertebrate Zoology 1907-1939. *Social Studies of Science*, 19, 387–420.

Stiegler, B. (2021). *Nietzsche et la vie. Une nouvelle histoire de la philosophie*. Galimard, Paris.

Tebes, J.K., Irish, J.T., Vasquez, M.J.P., Perkins, D.V. (2004). Cognitive transformation as a marker of resilience. *Substance Use & Misuse*, 39, 769–788.

Teglborg, A.C., Bonnafous-Boucher, M., Redien-Collot, R., Viala, C. (2013). L'innovation participative à orientation entrepreneuriale, un atout concurrentiel à cultiver. *Entreprendre et Innover*, 2, 23–30.

Thomson Klein, J. (2010). A taxonomy of interdisciplinarity. In *Oxford Handbook of Interdisciplinarity*, Frodeman, R., Thomson Klein, J., Mitcham, C. (eds). Oxford University Press, Oxford.

de Turckheim, E., Hubert, B., Messéan, A. (2009). *Concevoir et construire la décision. Démarches en agriculture, agroalimentaire et espace rural*. Quæ, Versailles.

Unsworth, K. (2001). Unpacking creativity. *Academy of Management Review*, 26, 289–297.

Utela, B., Storti, D., Anderson, R., Gautir, M. (2008). A review of process development steps for new material system in three dimensions printing. *Journal of Manufacturing Processes*, 10, 96–104.

Valéry, P. (1957). *Œuvres*. Gallimard, Paris.

Van De Ven, A.H. (1986). Central problems in the management of innovation. *Management Science*, 32, 590–607.

Vandenbosch, B., Saatcioglu, A., Fay, S. (2006). Idea management: A systemic view. *Journal of Management Studies*, 43, 259–288.

Vandervert, L., Schimpf, P., Liu, H. (2007). How working memory and the cerebellum collaborate to produce creativity and innovation. *Creativity Research Journal*, 19, 1–19.

Varenne, F. (2012). *Théorie, réalité, modèle. Epistémologie des théories et des modèles face au réalisme dans les sciences*. Éditions Matériologiques, Paris.

Véron, E. (1979). Dictionnaire des idées non reçues. *Connexions*, 27, 125–143.

Vigezzi, M. (2019). Les inventions partagées et les innovations concurrentielles : des approches nouvelles. *Innovations*, 59, 243–260.

Vilkas, C. (2009). Des pairs aux experts : l'émergence d'un nouveau management de la recherche scientifique ? *Cahiers Internationaux de Sociologie*, 126, 61–79.

Vinck, D. (1995). *Sociologie des Sciences*. Armand Colin, Paris.

Wallas, G. (1926). *The Art of Thought*. Jonathan Cape, London.

Walter, A., Helgenberger, S., Wiek, A., Scholz, R.W. (2007). Measuring societal effects of transdisciplinary research projects: Design and application of an evaluation method. *Evaluation of Program Planning*, 30, 325–338.

Wang, L., Xiong, C., Hu, J., Wang, X., Qiu, Z. (2018). Sequential multidisciplinary design optimization and reliability analysis under interval uncertainty. *Aerospace Sciences and Technology*, 80, 508–519.

Wang, L., Zhao, X., Wu, Z., Chen, W. (2022). Evidence theory-based reliability optimization for cross-scale topological structures with global stress, local displacement, and micromanufacturing constraints. *Structural and Multidisciplinary Optimization*, 65, 23.

Ward, T.B., Smith, S.M., Finke, R.A. (1999). Creative cognition. In *Handbook of Creativity*, Sternberg, R.J. (ed.). Cambridge University Press, Cambridge.

Weil, S. (1958). *Oppression and Liberty*. Routledge and Paul, Oxford.

Wen, M., Han, Q., Yang, Y., Kang, R. (2017). Uncertain optimization model for multi-echelon spare parts supply system. *Applied Soft Computing*, 56, 646–654.

Wolton, D. (2013). Conclusion. Pour un manifeste de l'indiscipline. In *Le gouvernement des technosciences ; gouverner le progrès et ses dégâts depuis 1945*, Pestre, D. (ed.). La Découverte, Paris.

Zeng, Z., Wen, M., Kang, R. (2013). Belief reliability: A new metrics for products' reliability. *Fuzzy Optimization Decision Making*, 12, 15–27.

Zhou, S., Zhang, J., Zhang, Q., Huang, Y., Wen, M. (2022). Uncertainty theory-based structural reliability analysis and design optimization under epistemic uncertainty. *Applied Sciences*, 12, 2846.

2

From Invention to Innovation

2.1. Preamble

> Individuals draw on information and keep only that which is relevant to their system of thought: they seek to reinforce certainties already acquired in the past. The processes of anchoring and objectification account for this necessity (Orfali 2005).
>
> When a team puts forward arguments designed to seduce funders or the public, but which the members and their peers know to be empty, the rule is silence. The citizens of the 'Republic of Science' know that, in a self-interested society, such mechanisms of disinterest are 'unfortunately necessary' (Stengers and Bensaude-Vincent 2003).
>
> The daily life of individuals and organizations is made up of mediocre decisions, because it's difficult to agree and work together, human intelligence has its limits, and hazards are inevitable. Mediocre decisions are the result of action that revolves around the objective (Morel 2002).

The first theory of innovation came from French sociologist Gabriel Tarde at the end of the 19th century (Tarde 2001). After a proof of concept (POC), a patent, etc., other processes are at work, with co-design or redesign of the project to create an industrial demonstrator (ID) development of the project (market "understanding" must be present at a very early stage), followed by the launch of production and marketing. The latter is a process in which technical possibilities and the market are constantly adjusted. But Christensen (1997) points out that when a disruptive innovation (DI) appears on an adjacent market (patents) and is the subject of a

---

For a color version of all the figures in this chapter, see www.iste.co.uk/andre/knowledge1.zip.

communication, an incumbent operator may create an autonomous unit and entrust it with developing and marketing the new innovation.

The year 2022 was shaping up to be a complicated one for GEAB 161 (2022) and for the West. Some thought that the pandemic would eradicate on its own and that things would return to normal. However, a mental confinement inherent in the health technostructure, highlighted by its catastrophic effects in times of crisis, is now infiltrating all scientific fields. The "old" economic system is cracking, but resisting despite difficulties linked in particular to economic recovery, and recently to the war between Russia and Ukraine. "Even if we know that it [the future] rests on the good use we make of new technologies, for the time being, in the absence of a realistic vision and political capacity, the gigantic sums of money released by states to finance this future tend to get lost: armaments, corruption [...]" (Plagnol 2020).

Disasters have an impact on fundamental aspects of society: revisited well-being, revival innovations, physical infrastructure, social connectivity, etc. Thus, the recovery of a shaken economy should result in a period of recovery, with an influx of resources, support for innovation (if only to correct emerging problems) and an accompanying synchronization of planning processes. It is, in fact, an opportunity to reorganize physical living and production environments by reducing the vulnerability of populations to future hazards. The aim is normally to have healthier, more resilient and more sustainable communities. However, in NAP (2015), the authors of this report have in the past found that efforts have tended to focus on essential infrastructure. In their view, few communities are taking a forward-looking, synergistic approach; as a result, important opportunities are being missed.

Furthermore, Whittaker et al. (2020) speak of the curse of countries with intermediate GDP per capita: their industrial production and their standard of living do not increase (unless they resort to borrowing and the public deficit). This middle-income trap that we find in France is located between countries producing at low cost and others that know how to exploit their innovations, if possible disruptive ones. However, according to Akcigit et al. (2017), states that innovate have accelerated growth relative to others. To achieve acceptable growth, it is certainly necessary to have the support of the state "in the right places" and at the right time (including research and training), but also entrepreneurs (or at least a pioneering spirit), supported by an enlightened tax administration (Aghion et al. 2020). The mastery of time plays a role in development and this historization emerges from speeches and actions which, for example, make institutions that protect private initiative and innovation the sine qua condition of success (Acemoglu and Robinson 2012).

So, the power of creative destruction identified by Schumpeter and that of the transformation of capitalism can steer society toward more sustainable, better-shared material prosperity. It must be part of a competitive drive by strongly supporting innovation (Rosa 2018). The concept of creative destruction is a process that implies that new innovations, rendering certain technologies obsolete, enable the creation of companies to compete with those already in place. So, overall, new jobs, new activities and new products are created and replace those that previously existed. To stay in the race, we have to accept innovation as a way of integrating a complex dynamic based on creativity, innovation, training, entrepreneurs and a state that supports the necessary fluidity. In this way, policy stretch between objectives that are difficult to explain and, in practice, to reconcile.

But to add to this, according to Babeau (2022), president of the "Sapiens" think tank, France is not in a position to be at the head of the pack of countries committed to this renewal, because of the weakness of its research and R&D. He writes:

> Abandoning an ambitious R&D policy has damaged our scientific habitus. Between a drop in the level of understanding of students and adults, a loss of confidence in science and a decline in the culture of risk, our country is no longer a scientific nation.

Israel finances 53% of its national research through support and joint actions from 300 foreign companies; Germany, through its Fraunhofer institutes, has contractual links with foreign companies, providing one-third of their budget, but in France, only 8% of the R&D effort is supported by foreign companies. Yet, domestic research does not reach the expected 3% of national GDP (2.1–2.2%), which discredits the prevailing view of France as a start-up nation, and raises the question of how to go beyond inventions alone (which, as Chapter 1 has shown, have their own problems)! According to Nature Index (2021), the number of publications in France fell by 4.7% between 2019 and 2020 (a bad omen!).

NOTE.– Trust:

Figure 2.1 (Saravi and Kordlouie 2009) brings together elements of the citizens' environment that reinforce their trust, particularly toward the state.

France Stratégie's "Les Métiers en 2030" report (2022b) is based on macroeconomic scenarios that take into account the above-mentioned developments, but smooth them out: the reference macroeconomic scenario is essentially based on evolutions, and not on major disruptions. In fact, in one of the figures in this report (see Figure 2.2), it is not envisaged that new professions linked to radical innovations will emerge, apart of IT.

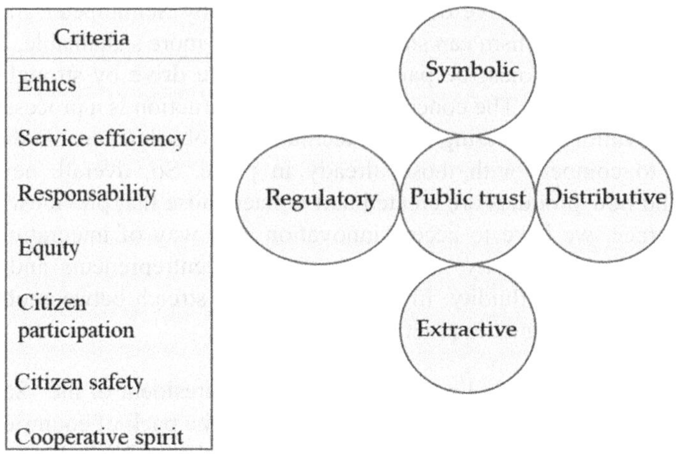

Figure 2.1. *Element of public trust in government*

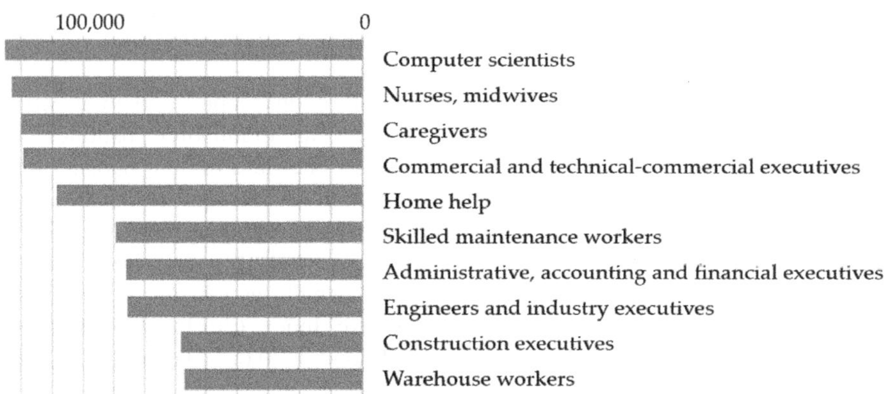

Figure 2.2. *High-growth occupations between 2019 and 2030 (right: % change over the period)*

For example, Forest et al. (2011) had already sought to move away from the model of engineering training based on mastery of essentially technical problem-solving. The authors envision training that is more oriented toward innovation, addressing complexity and its holistic treatment (with the support of the human and social sciences).

> The human and social sciences are no longer there as a soul supplement or a tool for professional integration, but rather as an

integral part of the engineer's work, at the very heart of his or her profession. Where are these recommendations in the above-mentioned report?

NOTE.– Pandemic and its consequences:

For around 2 years now, we have been suffering the effects of the SARS-CoV-2 (Covid-19) pandemic. Distraught politicians have switched from talking about an ordinary flu, with no major effects, to more credible prevention obligations, sometimes with a delay. Today, France counts over 130,000 deaths! The whole of society (including scientists) has added its own vision and proposals to the prevailing vagueness, which has not helped matters. Using (for the most part) methods of solving classical differential equations, we find most of the connections between viruses (growth, exposure, disappearance) and individuals (sick, immunized, etc.), which have already been scientifically validated (Reversat and André 2022). Trends derived from this semi-quantitative modeling validate a number of effects: social distancing, barrier gestures, mask quality, the value of teleworking, etc., but envisage successive shifts.

The exceptional does not last. Production structures remobilize their forces for their own "benefit" when perhaps it would be better to continue mobilizing operational forces for the benefit of the community, or communities, or to take into account the organizational and creative achievements of those who played the game of generous aid during the crisis. Between anarchic solidarity, national dirigisme, egoism and inertia, there may be a way of finding real legitimacy without being obliged to give "good points" or a nice medal to those who wish to participate in the national effort.

However, if we are to find a new social agreement for the future, taking into account the many outbursts of solidarity that have been observed, it is imperative that we remain present and continue to manage the health and socioeconomic crises, but undoubtedly in other ways than in this emergency, which is ultimately stimulating because it generates links of all kinds. What we need to do, then, is to embark as quickly as possible on a process of reflection and proposals, underpinned by "societal" considerations, aimed at preparing and changing tomorrow, which should not be limited to business as usual, a return to the old intangible dogmas, virtually elevated to the rank of ethics where we need a human as unusual. But this is undoubtedly more difficult to do than to say. The state will undoubtedly not escape the proposal for a new pact for the future involving a new impetus, societal policies redesigned in the light of alternative values to be reaffirmed, a more flexible, agile, efficient and coherent, prioritized internal and external solidarity values, other decision-making methods. There should be a place in this "tsunami of recovery" for those who have been day-to-day players in this great adventure (for those who have not been affected directly or indirectly by the virus).

Another example that may be of interest to citizens and scientists alike is that the constraints put in place to flatten the epidemic curve have limited economic opportunities and exacerbated the recent recession, because we now live in economies with highly fragmented production systems. Production is largely just-in-time. This mode of profitable production, linked to the specialization of companies and countries, makes the economy vulnerable to major disruptions, since a production stoppage at a single key company can paralyze the entire value chain. We can experiment calmly within a framework of social trust to make progress on these issues. But, to achieve such objectives, the state could abandon its bureaucratic ponderousness in favor of a better integration of civil society in its actions, enabling the emergence of forward-looking ideas through trusting dialogue (André 2020).

The strategy for most developed countries during the pandemic was to implement containment measures. Such measures, necessary to limit the mortality rate of the disease, have a significant economic cost. The constraints put in place to flatten the epidemic curve limit economic opportunities and exacerbate the coming recession, because today we live in economies whose productive systems are fragmented production systems. Production is largely just-in-time. This mode of profitable production, linked to the specialization of companies and countries, makes the economy vulnerable to major disruptions, since a production stoppage at a single key company can paralyze the entire value chain (Gerschel et al. 2020). The war between two European countries reveals other vulnerabilities that need to be addressed.

Thus, between a "great controlled slowdown" facilitated by rising interest rates and the growing critique of the dynamics and natures of progress, with the usual mix of autonomous crises and reactions leading to Europe managing to find its own way (perhaps one that would pave the way for the reinvention of global governance), GEAB 161 (2022) considers that we are in the midst of a double crisis dynamic, that of the world before and that of the world after, whose fundamentals should undergo a major overhaul.

> It's a real "moratoire on the future" that we are about to experience. It will undoubtedly give the impression of a return to the old system, which, exhausted, will find it hard to resist the weight of a society impatiently chomping at the gates of its future (new generations, economic players or 'reformers' of all stripes).

This significant and burdensome uncertainty, the lack of financial support and the risks of business as usual are therefore a major obstacle to innovation (and for the invention that underpins it), a new "valley of death"! But, as Daloz et al. (2020) point out, "shaping the unknown is not necessarily 'worse' in terms of risk and financing. While an increase in uncertainty may encourage investors to become gamblers, facing the unknown requires investors to understand the design logic and adopt a perspective on the new potentialities to be explored. If markets and technologies are unknown, good design is not about more risky trials – it is about designing technologies and markets to match a winning lottery.

Valley of death no. 8: existing and being supported by society as a whole

2.2. Introduction

> It seems that for health at any price, big business has taken our money, the authorities have restricted our freedoms and politics has captured science (Biron 2022).

> Orwell and Huxley made an analysis suggesting that Man reduced to its means alone can only improve by renouncing its freedom, a freedom that the human sciences, at the same time, presented as an illusion (Dufresne 2022).

> Beings as 'uncultured' as peasants and workers often possessed an understanding of beings and things, a finesse of mind and an openness of heart, in short, a wisdom that far surpassed, in delicacy and depth, that of the average intellectual (Thibon 2022).

Hohberger (2016) shows that patents with intermediate levels of scientific intensity have the highest propensity to be cited; however, the fastest rate of diffusion is observed in patents with high levels of scientific intensity. For innovation, it can be advantageous to play on this register to attract industrialists, but the existence of a POC for a given invention is essential because it demonstrates the technical feasibility of an idea (which is not always the case with a patent).

The word innovation has become almost overused because, thanks to the media, it seems to be everywhere. Yet, in reality, new products and services that improve on old ones emerge from rare radical innovations. Let us remember that the concept of innovation (which covers the fields presented in Figure 2.3 according to Blankesteijn et al. (2021)) is as follows: "The process of translating an idea or invention into a good or service that creates value with customers willing to pay for it". In many minds, this is still based on the "old" leitmotif of the 1933 Chicago World's Fair: science invents, industry produces and the public buys.

The term "satisficing solutions" (Simon 1976) is used to describe those with which we are satisfied, for lack of being able, in practice, to determine an optimal solution corresponds more or less, for the moment, to what research can propose (logic of offer or bottom-up). This linear model has been much studied, and naturally criticized, over the last few decades (Godin 2006; Balconi et al. 2010), as it does not envisage a demand-driven approach. And yet, while growing awareness of the issues at stake is present at all levels, whether institutional, entrepreneurial or societal, the operationalization of change is still struggling, despite the fact that societies are geared toward innovation (Chouteau et al. 2020).

Research	Development	Demonstration	Deployment	Improvement

Figure 2.3. *Principle of transfer (linear model)*

In this chapter, after moving from an idea transformed into a POC on the basis of Figure 2.4, where creation is based on short timeframes, with a range of possibilities depending on people's experience, the aim is to examine how we can reach society through innovation. At the same time, the transition to the "other side of the looking glass" is reflected in developments in the socioeconomic and/or social fields, as shown in Figure 2.5.

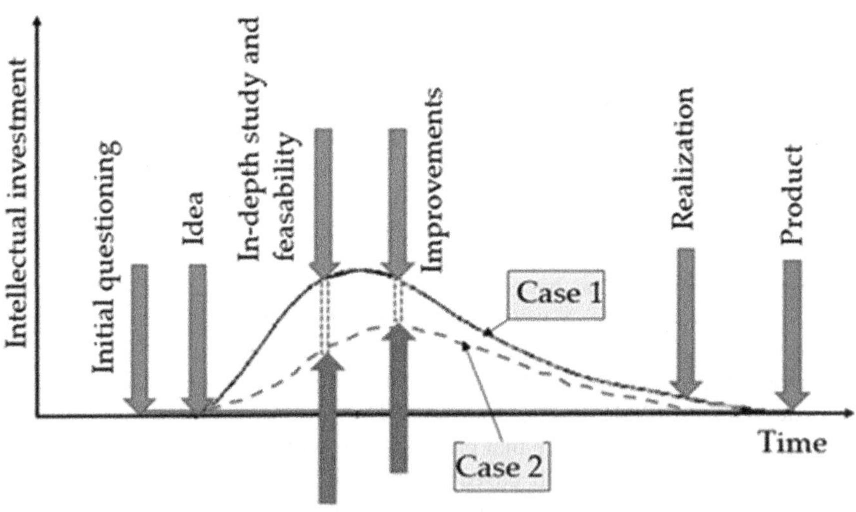

Figure 2.4. *Example of an innovative process from idea to product with two possible examples (case 1: high intellectual investment; case 2: more modest)*

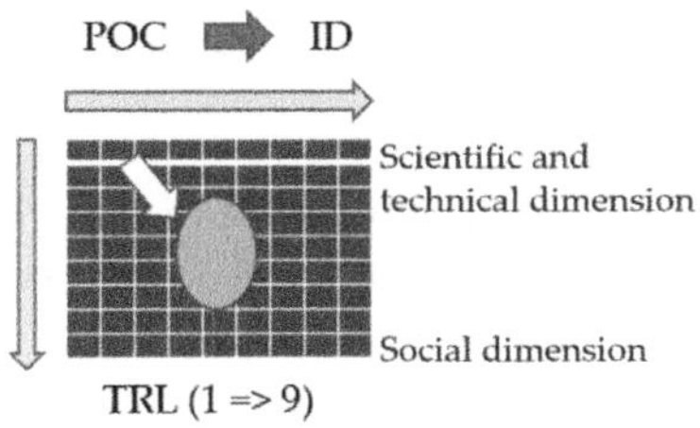

Figure 2.5. *A transition from POC technology to industrial demonstrator (ID)*

The origin of innovation comes either from "incremental" creativity or "divergent" creativity. Incremental creativity concerns improving existing knowledge and ideas, while divergent creativity is about breaking with existing ideas and skills. It is therefore useful for exploration (with real risk-taking) and the former for exploitation ("prudential exploration", according to Le Masson et al. (2006)). Divergent creativity at the origin of disruptions leads to the emergence of original associations between already existing knowledge. At the breakthrough invention stage, the potential users are not all known, nor are the means of reaching them.

In such cases, it can be advantageous to combine technological knowledge with other, "real-world" knowledge, to obtain less mutilated intermediate results (Garel and Rosier 2007), a kind of "semi-finished product" (Weil 1999). According to Garel and Rosier (2007), this term covers innovation situations that are divergent innovation with strong expansion in terms of technology and value. In the end, it is when the following elements meet: customer or user needs, technical solution and corporate strategy. At the invention stage, the ways in which innovation can contribute to economic development are essentially the same, and this is what the presentations in this chapter are all about.

Thus, it is a question of trying to better understand how we connect invention and innovation. Innovation is therefore, unlike invention (see Chapter 1 of Volume 1), an activity of establishing relations with society: the identification of the technology is not directly linked to the invention, but to a successful invention. It requires, within the current determinism, the identification of a need or a potential market for a new product or a new process (Freeman 1988). This means that we are able to propose "well-defined and transferable processes intended to produce certain results deemed useful" (Friedman 1952). This context therefore requires the definition of clear methodologies on the modes of operation relative to an application end. The translation of possibilities then aims at scientific reductions while ensuring production and use. New innovative proposals should be considered as expressing several key general concepts: initiative, creativity, responsibility, trust, meaning, sharing, process, flexibility, culture, heuristics, learning, etc. These areas contribute to adding complexity to the problems raised in terms of innovations (advantageously disruptive).

As has already been reported, Maslow (1966) wrote that a hammer tends to see all problems as a nail! The creativity that is often put forward with its many clichés or images of Épinal constitutes a necessary starting point for innovation, in particular, that resulting from research exploring teleological approaches, because it forces us to think differently within increasingly complex and uncertain systems. For the authorities, this situation has become urgent, without being new in substance. It induces a new phenomenon in research structures, in particular academic ones, that of taking the risk of failure (but in the event of success, whoever arrives first wins everything, especially if they are already recognized by their peers and its hierarchy), which leads to a certain conformism, translated by a number of filters that the author has spotted under the radical name of "valley of death". But, as in Orwell's *Animal Farm*, some are more equal than others.

A major "valley of death" corresponding to this transition from POC to demonstrator exists and needs to be clarified. The framework for possible development through external support (personnel, funding) is based on the importance of mutual understanding, sense-making and consensus-building between internal partners (especially if interdisciplinarity is action) and externally.

Kijkuit and van den Ende (2007) consider that DIs are focused on the structure and content of different networks, and emphasize the need for strong links and related prior knowledge, the integration of decision-makers and the transition over time from a large, nonredundant and heterogeneous network structure to a smaller, more cohesive network structure capable of efficiency (see also Reiter-Palmon and Illies (2004)). A bill proposed by two members of parliament aims to change all this:

> France needs more capital for its companies at a time when the digital revolution is shaking up the organization and our lifestyles. To develop industry, promote the ecological transition and liberate our energies, we will provide our companies with the means to fulfill their ambitions. In order to achieve this common goal, the present bill aims to create an *Innovation Booklet* (Assemblée Nationale 2022).

Following on from the difficulties outlined in Chapter 1 of this volume, it may be possible to resolve at least one of them, that is, financing (but we do not know how or at what level).

Innovation from invention follows a complex process that leaves plenty of room for improvisation and chance: idea, the possibility of initiating an in-depth project, proof-of-concept studies that take into account relations with society (co-creation), market research, financing plans, business plans, initial financing, creation or integration into a suitable organizational structure, ID and its possible detour, canvassing of new financial partners, additional fund-raising, prototyping and market testing, manufacturing, launch, etc. An invention is not enough, as there is a long road between the idea and its realization (Merle 2019). Figure 2.6, taken in principle from Harryson et al. (2008) on the one hand, and Kotze and Purgathofer (2007) on the other hand, shows how we move from a "happy" idea validated by a POC to a profitable application.

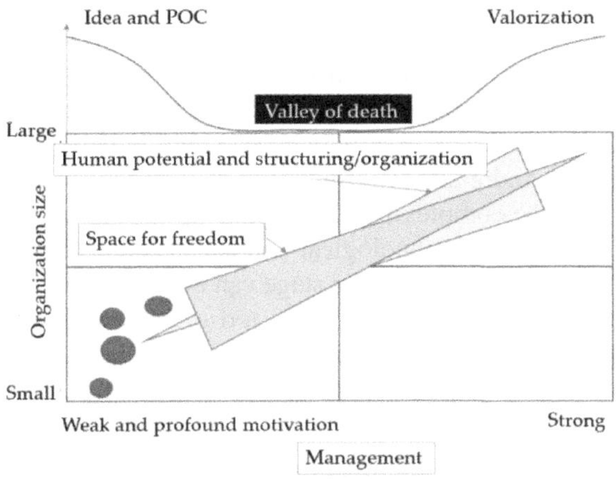

Figure 2.6. *From idea to application*

NOTE.– Patents and applications:

According to the 2015 China Patent Survey Report published by the State Intellectual Property Office of China (SIPO), the average implementation rate of invention patents is only 50.9%, indicating low conversion rates. Furthermore, in 2014, only 12.6% (286,488 out of 2,265,224) of all utility model patents were maintained for more than 5 years, compared with a global average of 18.2% (Zhou and Wang 2020). Patents are a good indicator of future applications, but the transfer yield is not equal to one!

Co-creation initiatives that involve civil society serve to disseminate inventions in a number of ways. By including users (or their close contacts), they can help guide the transformation of a technical POC into an innovation demonstrator (ID) that meets the most relevant and appropriate needs and facilitates future adoption. However, for co-creation initiatives to materialize successfully, four challenges must be overcome, according to OECD (2021b):

– Cooperate with key stakeholders to build a shared vision, provide clear incentives and set milestones to increase visibility of progress and motivate stakeholders. Stakeholder engagement is difficult in the presence of asymmetries and cumbersome hierarchies.

– Adopt effective and traceable governance and management systems.

– Agree on the ownership and use of jointly developed data and intellectual property.

– Adapt to changing internal and external institutional environments.

But on the frontiers of science and technology are ideas (brick-red ovals) that increasingly require the mastery of complexity (changes in spatial and temporal scales (Dodet 2001; Puech 2016)), which we know will require interdisciplinary convergence actions (André 2017c). If interdisciplinarity is a means, can we propose approaches to tackle the complexity of situations in which the innovator alone must imperatively find scientific, technological and socioeconomic alliances for his idea to succeed? The role of foresight at this stage can be of great interest (as long as it is carried out in the right way and not, as is too often the case, to support certain lobbies). The same applies to the strategy to support innovation which must support creativity, interdisciplinary project management, in short, risk-taking in action, managing lakes of uncertainty and lack of knowledge (see heuristic approach), finding the paths (if any) to reasonable success.

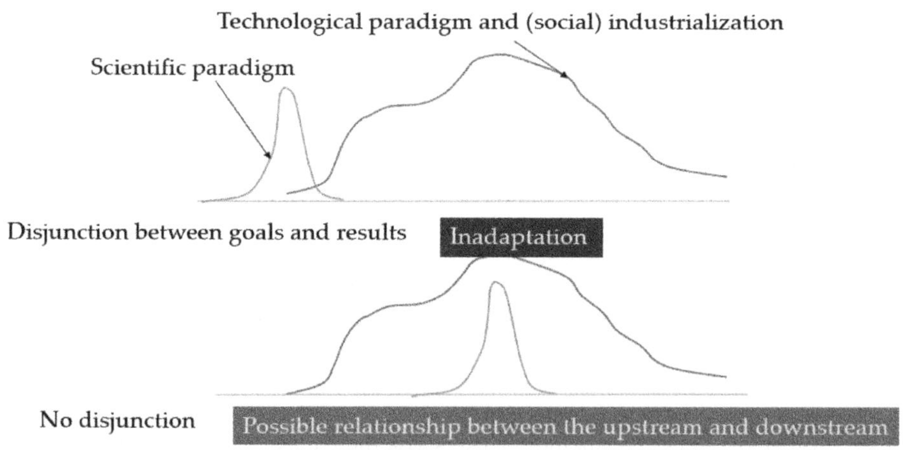

Figure 2.7. *The transition from an invention "outside" the need, requiring a closer relationship with the latter in order to be taken into account*

In any case, it is undoubtedly necessary for inventors from the "hard sciences" to move beyond the "artisanal" aspects they have been practicing (and which are scientifically and technically useful) to try and convince their national and international scientific communities. However, this consideration, which can amplify a movement, is only a subsidiary contribution to the project, which aims at application. This means directing activities in directions of social (or societal) utility, integrating directions/goals that the players in these disciplines have not thought of. It is a matter of learning more optimal co-construction, as defined in Figure 2.7.

Maslow (1954) showed that human motivations are born of needs to be satisfied, hierarchically divided into five main levels, as shown in Figure 2.8, which represents Maslow's pyramid, from physiological needs to the ultimate goal of self-fulfillment. These needs create tensions that generate motivations. We can certainly invent for ourselves, but if we are preoccupied with industrial applications, it is a good idea to keep these precepts in mind.

Figure 2.8. *Maslow's pyramid*

However, to innovate with the environment, the following rules can be taken into consideration:

– how the operation is built can be just as important as what is built;

– identifying stakeholders in advance;

– maintaining the trust of employees, who may fear that new practices could cost them their jobs;

– starting small (but do not stop thinking big). Not every project has to be gigantic, vast or grandiose;

– knowing where your "sand pile" stands in relation to others, and setting clear objectives;

– innovation generates more innovation; communication about innovation activities is essential;

– for sustained support, sponsorship from senior management is essential; or the necessary funding must be available;

– commitment to a transparent and critical examination of successes and failures.

However, Aggeri (2020) warns us:

> Just like a street lamp on a dark street, decision-makers in both the public and private sectors make their decisions according to the areas illuminated by the instruments at their disposal. To understand actual decision-making processes, and not just strategic intentions, we need to take a closer look at management technologies, a veritable "invisible technology", i.e., the underpinnings of organized action.

At the same time, Alsaaty and Sawyer (2019) sensibly suggest:

– distinguishing between key and minor project trends;

– identifying problems from the end-user's perspective;

– prioritizing and evaluating solutions;

– identifying needs based on existing capabilities and end-user requirements;

– finding the necessary partners and their roles;

– identifying the activities required to implement the desired strategy.

2.3. Methodologies to be put in place

> Leaders seem to have been trained, even formatted, in a world of realist thinking, which leads them to apprehend the world as something external to them: this intellectual heritage [...] seems, in fact, to prevent or at least slow down the apprehension of this complexity [...], constituting a kind of realist filter (De Geuser and Fiol 2003).

> We are on the threshold of a new cultural compromise of culture, in which traditional humanism will have to settle its affairs with the third culture it has never acknowledged. And the third culture is neither the hard sciences nor the humanities, but all the knowledge of engineers (Sloterdijk 2000).

> We are all "guilty" and "responsible" for the situations we live in, despite ourselves, because we didn't choose them or build them. We are responsible because this structural "equality" is not a matter for an elite to decide the point of being of the situation, it's all of us (Benasayag 1994).

"The old scientific ideal of *epistēmē* – of absolutely certain, demonstrable knowledge – has proved to be an idol" (Popper 2002). By bringing together the so-called "hard" sciences (invention) and the human and social sciences (HSS), we need to master the links between actionable knowledge and general theories in management science (and relationships), and in particular to know how to make scientific results more actionable. Creating actionable knowledge, making theories more actionable and bridging the gap between technology and practice are just some of the issues involved. The research to address these issues are collaborative (Tsui 2007), working with organizations that are pioneers or willing to experiment with new ways of organizing themselves by working in the field, with partner companies and organizations, to better formulate and solve innovation and change management problems. The idea is for HSS research to produce applied/applicable knowledge (concepts, models, tools, methods and analytical grids), but from a practical point of view, to exploit interdisciplinary interaction for innovation.

Questioning disruptive inventions is a good "candidate" for this kind of thinking. This can have two components: one aiming for a science-technology-society resonance, where the invention is presented with an applicative goal that can be used in the short-to-medium term; the other where, in a more distant sense, the invention is merely a means serving as a starting point for exploring applicative potential in other promising fields. This "offering" activity can be confronted with knowledge of what is real and what is possible, using multi-criteria decision support. Decision-aid tools lie at the intersection of several disciplines, such as psychology, sociology, management science, economics, law, but also scientific and technological knowledge (including, notably, the sciences of management innovation and technology management). We find ourselves once again in an almost familiar environment, that is, of interdisciplinarity.

In this context, it is important to overcome breaks with classical operational research (Taylor 2008) in management: limits of decision models with single rationality and difficulties linked to the activity of optimizing decisions. In doing so, the development of a three-pillar paradigm, consideration of multiple criteria for decision support, preference for reasonable and robust solutions rather than strictly rational, and conception of the aid relationship, in decision support, based on the search for consistency with values rather than on the search for an optimum can make it possible to identify operational trends on the two interdependent axes inscribed in the medium and longest term.

Valley of death no. 9: access to co-creation

Figures 2.9(a)–(c), taken from Fanvoice (2021) and corresponding to work involving co-invention, show the many interests of this practice, which largely answers several research questions that can be considered when dealing with proposals/prescriptions for change, such as:

– the optimal nature of the (lay) co-creation group(s) (Naudon 2013; Deserti et al. 2022);

– the necessary steps in the co-creation process within a framework of social responsibility (Wittrock et al. 2021; Chapter 1 of Volume 2) based on the construction of a mental representation at the interface of disciplines (Bourdin 2022);

– the construction of decision-making processes of group(s) involving actors with different roles, skills, knowledge and expertise (according to Jakobsen et al. (2019), conceptualize multidimensional approaches linking the scientific, material, discursive and organizational dimensions of innovation and their synergistic effects on major societal challenges);

– the search for the mode(s) of representation of the invention and the diversity points of view;

– the key role of public representatives in the various stages of the co-creation process.

Activities are not solely focused on technical innovation, but can also include crowdsourcing and the extension of action research activities to a wide range of audiences (schoolchildren, curious citizens, senior citizens and other vulnerable groups, various associations, trade unions, elected representatives, etc.). The projects cover a wide range of environmental issues.

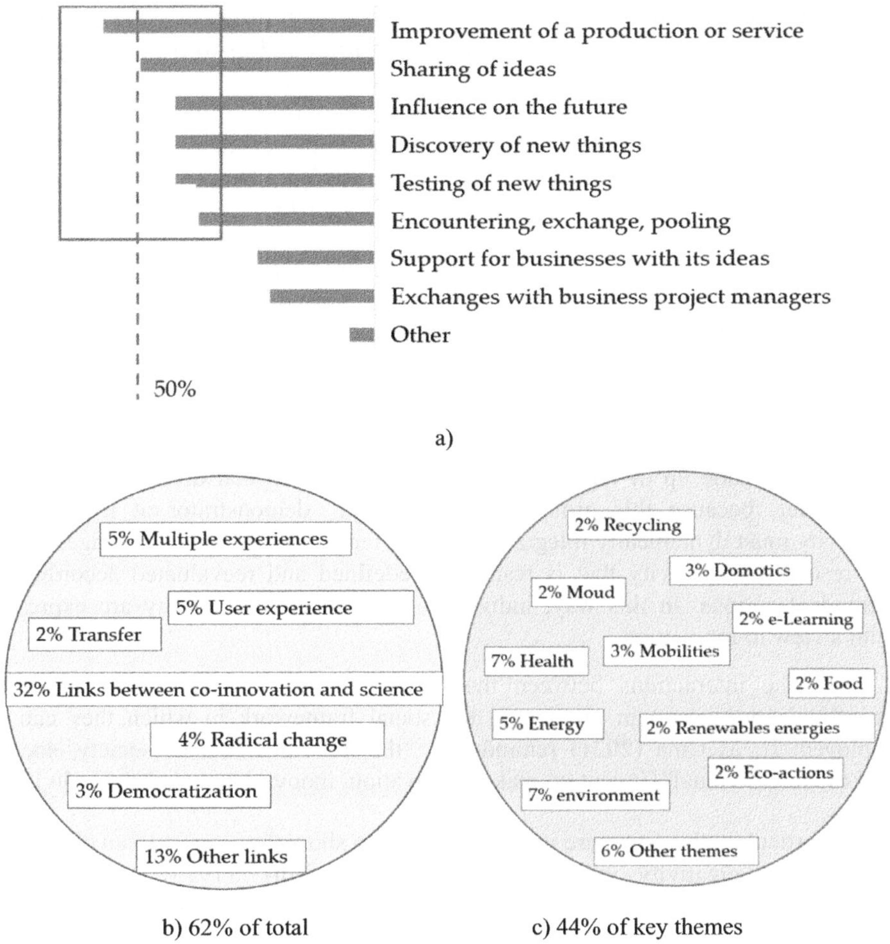

Figure 2.9. *(a) Benefits of co-creation; (b) links with science; (c) key themes*

From a practical point of view, co-creations make it possible to master the way in which the dynamics of multidimensional and multi-scale innovation can be initiated and followed to envisage the creation of economic value (with solutions to environmental and social challenges in parallel). However, it is useful to test the proposals to ensure the relevance of the final product. To address credibility issues, transparency in the process, including prerecording of assumptions and filing of an action plan before execution, should eliminate the temptation to report the best results after the completion of the project or to revise the assumptions once the results of the study are known (Tsui 2016).

The idea of co-innovation is to enable a group with different skills and from a variety of fields to discuss and try to bring one or more ideas to fruition, promoting the democratization of emerging proposals and involving society in scientific and technological issues (with concrete implementation in everyday life). This joint project also draws on the notion of the imaginary in management science (De Cock et al. 2013; Komporozos-Athanasiou and Fotaki 2015; Le Du 2017; Chemin-Bouzir 2019) and some others that may be related to it (Vygotsky 1999).

In this context, this imaginary can help us to understand the context of creativity in organizations associated with a broadening of technological through co-creation with partners connected to society, which depends on:

– History and values of the organization, as was the case for invention.

– "Personal" considerations of the individual working there (*idem*).

– Group made up of representatives from disciplines that do not talk much to each other, because this group has a "physical" demonstrator at its disposal, creativity must dynamically integrate what is effectively separate knowledge, and be the result of an activity that is reshaped, redefined and reevaluated according to individual actions. In this way, individual and collective creativity are expressed with a view to innovation.

– Of the interactions between these elements, of the understanding of the invention's purpose(s) in relation to the social framework in which they can be deployed. If, as Cina (2021) reminds us, "the individual makes society, society makes the individual", invention makes innovation, innovation makes the individual.

We articulate this literature in such a way as to show the interest of an alternative approach to creativity management using Castoriadis' (1975) notion of the imaginary. This enables us to understand the creativity of an organization without overlooking the fact that, like the individuals who make it up and the tools it employs, it is historically and socially situated. In this way, it enables a holistic approach: it encompasses the social, technical, organizational and individual aspects of creative development, without forgetting that these contextual elements are constantly evolving, at the same pace as the individuals who work.

A Living-Lab or living laboratory (Bulkeley et al. 2016), understood as a site for designing, testing and learning innovation processes in real time, reduces the uncertainties associated with the new, in order to respond to particular societal, economic and environmental challenges. This passage "to the real", offered in such a space for negotiating definitions and understandings of problems, various claims on resources, authority or dominant ideologies, the laboratory can translate into changes in innovation practices.

In stable situations, results are accumulated and publications are subject to known regulatory mechanisms. However, when these are no longer functional, the disruption caused by divergent revisiting imposes itself as an element of disruption for a possible way out of impasses. As Edgar Morin explains (quoted by Bouiss (2020)): "Crisis depends on randomness: at certain crossroads, it is possible for a minority, for an individual action, to tip development in a direction that is sometimes highly improbable". For the "engineering sciences" component at the origin of an invention (which has passed several filters), the first obvious thing to do is to demonstrate the need to orient research toward the need (we are dealing with a technology here); on the other hand, the expression of possible applications must lead to top-down or teleological approaches, or, the opposite problem, to project management that seriously integrates society in an operational way, for "premature" options or returns to more mono-disciplinary scientific subjects.

2.4. Innovation policies

> Innovation is the adoption of a new practice by a community [...]. Success at innovation lies in the intersection among the innovator's domain expertise, social interaction skills and ability to recognize and move into opportunities (Deming and Dunham 2010).
>
> For bureaucrats and industrialists alike, the most important thing is that these "technologies of the future", which further threaten the natural environment, human health and privacy, enable them to consolidate their dominant social position and open up new profit prospects (Luzi 2014).
>
> In other words, a creator's entire effort [...] must be devoted, according to the very essence of his work, to creating enjoyments for the public that require little or no effort. It is from the audience that he must deduce what touches, moves, caresses, animates or delights the audience (Valéry 1957).

In France, with the ANR and, in Europe, with Horizon Europe, the aim is to produce the new (and predictable) through calls for projects, while at the same time pitting researchers and, more broadly, teams against each other. This policy is unequal, as it sidelines researchers and disciplines that could lead to progress for society and above all in the choice of directions to explore (albeit within a framework). The emphasis is on "research that counts" (but is predictable), which is beneficial in many situations.

According to Rau et al. (2018), disadvantages have also emerged, such as the increasing marginalization of basic research or "research for research", which may or may not have effects on direct applications. This unequal situation is reinforced by the labeling of Carnot institutes, IDEXs (initiatives for excellence), etc. (Harroche 2021). Aghion et al. (2021) consider that, in order to escape from "path dependency", it is advantageous to support "horizontal", non-targeted policies that will open up the spectrum of innovations to participate in an industrial reconquest:

> These horizontal policies include in particular: (1) better funding and governance of basic research (universities, Agence nationale de recherche, CNRS) and better governance of basic research (universities, Agence nationale de la recherche, CNRS); (2) encouragement of sponsorship; (3) a greater role for venture capital, private equity and institutional investors; (4) tax incentives for innovation; (5) flexicurity in the labor market to encourage creative destruction (Aghion et al. 2021).

> Segregation certainly defines the ghetto imperfectly, as it's a simple measure of population heterogeneity versus homogeneity that doesn't take into account how people cross the ghetto's material boundary. But relegation is no better, because keeping a person in an inferior position or rank does not necessarily imply spatial confinement. What's more, relegation refers to a mental process, and from a social semiology perspective, it's always preferable to replace an invisible process with observable clues. Stigmatization must not simply be associated 'in one way or another' with confinement; it must be the cause of ghettoization (and ghettoization must be a direct result of stigmatization) (Raynaud 2021).

With major societal challenges (including global warming), politicians are obliged to make choices, as clearly as possible, using a reductive approach. Simple and effective, it consists of focusing on a framework of key technologies that are already transforming the very foundations of our economy. We therefore support what is already visible, with often complicated incremental requirements. These conditions are normally comprehensible to the general public; the principles that are accessible therefore concern the main drivers and levers of systemic change. The various players involved in this "causal" progress seek to accelerate effects to avoid risks or secure benefits. The race for performance is likely to generate powerful feedbacks that will keep the momentum going.

NOTE.– Patent relations – fields of application:

Kelly et al. (2021) have proposed a criterion based on the analysis of patent statements that traces the history since 1840 of this relationship. Technological indices capture the evolution of technological waves, more or less sensitive for the public, as shown in Figure 2.10.

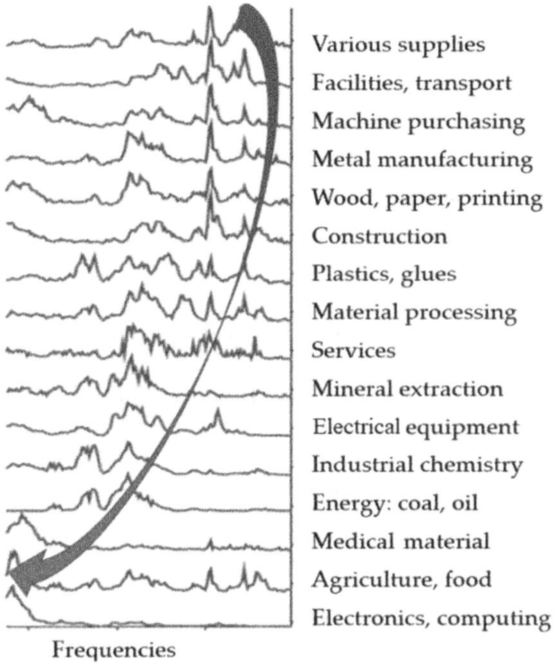

Figure 2.10. *Illustration of trail dependency (source: Kelly et al. 2021)*

"A disruption occurs when a new technology of equal or greater capability becomes available at a significantly lower cost than existing alternatives, after which the incumbent technologies are replaced very swiftly" (Arbib et al. 2021). The scale and impact of these emerging technologies, which were initially unsupported, are therefore largely underestimated. However, linear thinking is certainly not totally suited to finding solutions to problems linked to major societal challenges perhaps much more quickly and comprehensively than is usually thought of as possible. Conventional thinking may leave some creative people by the wayside, but that is just the way it is! Figure 2.11, taken from Arbib et al. (2021), highlights this competition between tradition and disruption.

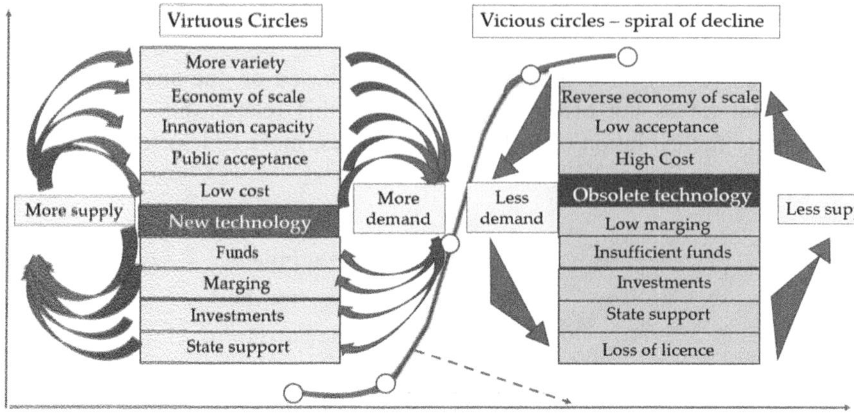

Figure 2.11. *Between vicious and virtuous circles in innovation*

According to the same author, technological disruptions are not the result of a linear process. Their evolution (provided they demonstrate their profitability) can be very rapid, with a very broad social impact. However, this emergence is complex:

– various obstacles can limit change;

– convergence to create new products and services;

– new artifacts begin to be deployed in a niche market before being extended to others;

– industrialization is driving down costs as capacities improve;

– adoption is nonlinear and follows an S-shaped curve;

– the obstacles fade with a virtuous cycle for new products and a vicious circle for old ones;

– local disruption is extended to the entire production system and value chains;

– the effects of disruption spread, transforming other sectors and society.

2.4.1. *An ambiguous situation*

> In social groups whose aims and meanings are as different as can be imagined, we find, despite everything, the same formal behaviors of individuals among themselves. Domination and subordination, competition, imitation, division of labor, party formation, representation, solidarity all this can be found in state society as well

> as in the religious community, in the gang of conspirators as well as in the grouping of economic interests, in the art school as well as in the family (Simmel 1999).

> What's epistemology for? Such a question calls for a provocative answer. Epistemology serves no purpose, if we take the word "serve" in a strictly pragmatic sense: epistemology has no direct effect on the advancement of science (Granger 1992).

> Civilizations that increase their capabilities and reach the fastest, spread through conquest or mimicry. Zero-sum competition forces societies to exploit or be exploited, hardwiring environmental degradation and inequality into the system (Arbib et al. 2021).

In the minds of national decision-makers, having ideas and knowing how to disseminate them to enable profitable innovations means "putting territories in competition by giving them virtuous conditions for economic development" (Casadella et al. 2017). The state may feel that it is positively translating this "creativity into skill-building" in the form of innovation policies. The European Union with its plans (Horizon Europe), like France (ANR (France Stratégie 2022a)), seems to have thought about what is essential for society.

Indeed, there are major trends that need to be addressed as a matter of priority (energy, depletion of reserves, global warming, ageing, health, migration, war, etc.) and which legitimately require appropriate treatment. The associated funding is sufficiently large to involve the research community in scientific studies aimed at incremental innovations that escape the principles of co-creation (see Figure 2.12, highlighting significant expenditure, but notable differences between countries according to the CNRS (2022)).

Indeed, apart from the MITI-CNRS (2022) proposals, which aim to integrate knowledge within the framework of interdisciplinary actions (poorly financed), we remain within the model, with the funding and implementation of major R&D programs. To the author's knowledge, there is no subsidiary complement to support a policy of research, innovation and creativity oriented toward the management of more radical knowledge and innovation systems. In addition to its role as a funder, the state should proactively implement measures to improve national and regional economic and technological attractiveness, outside the scope of the above-mentioned plans. Rodrik (2007) stresses that policies must support innovation through financial backing and by stimulating the dynamics of collective interaction.

For Sicard (2021), France's industrial decline reflects an economy in decline. For over 20 years, industries have been unable to sustain the economy, resulting in insufficient growth in GDP, very high unemployment, "public spending at the highest level of any OECD country, compulsory deductions at record levels, a trade balance that is still heavily in deficit, and the country's indebtedness rising inexorably year after year". For this author, industry is the engine of growth in a country (see Figure 2.13). So, a 10% increase in industrial production would result in a 6.4% increase in per capita wealth, and to return to a balanced public accounts situation, the industrial sector would have to be rebuilt to the level of around 18% of GDP.

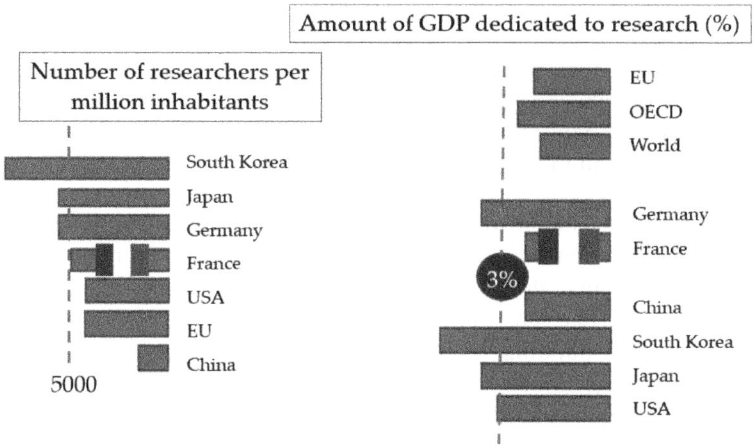

Figure 2.12. *Support for innovation*

While this figure is enlightening, other elements provided in Aghion et al. (2020) support this view based on solid data (see Figures 2.13 and 2.14). In this situation of retrenchment, the role of the state cannot be overlooked. Lewis (1980) writes: "The two conditions for sustainable growth in a country are, firstly, the availability of entrepreneurs and administrators and, secondly, the necessary fiscal capacity and savings"[1]. Schumpeter (1984a) reminds us that: "The fiscal history of a people is an essential part of its history. The fate of nations depends to a large extent on the economic sacrifices the country requires and the use to which those sacrifices are put". We need to find the ideal combination of economic stimulation through innovation and creative destruction, developed by Schumpeter.

Today, industry only contributes around 10% of GDP and France, excluding Greece, is the EU's most deindustrialized country (although, in 1960, this rate was around 26% of GDP). So, the loss of wealth associated with the decline of industry has forced public authorities to increase social spending and compulsory levies.

1 Author's translation from the French version of the text.

Clearly, innovation is certainly the way forward if our country is to regain the dynamism its citizens want!

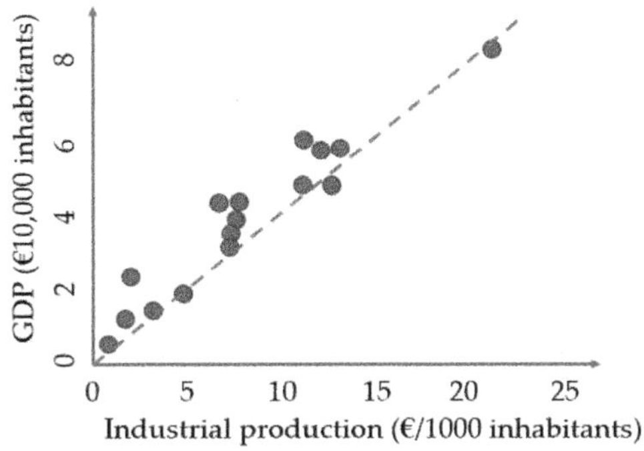

Figure 2.13. *Relationship between GDP versus industrial production*

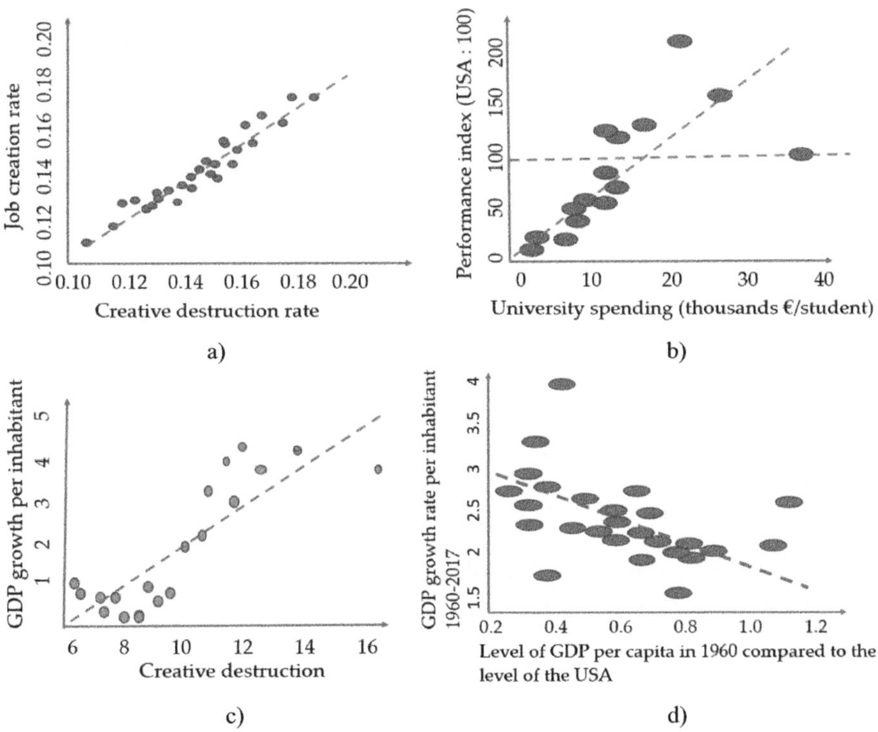

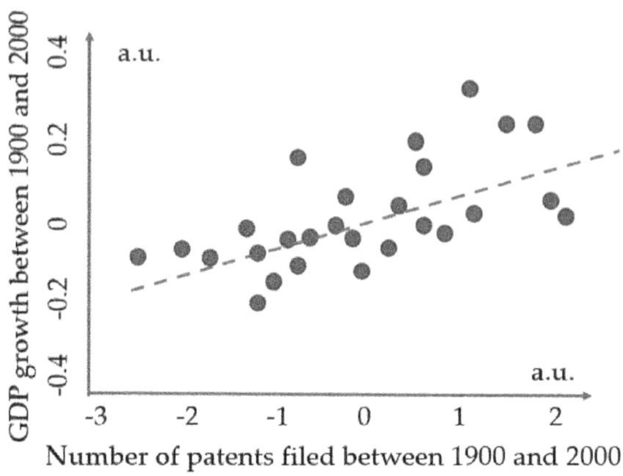

e)

Figure 2.14. *Relationships between economic power and innovation. (a) Relationship between job creation and destruction in the USA; (b) relationship between spending per student and Shanghai ranking; (c) correlation between growth in GDP per capita and rate of creative destruction; (d) average growth rate in GDP per capita between 1961 and 2017, as a function of standard of living in 1960; (e) correlation between growth in GDP per capita and patents*

2.4.2. *Technological paradigms*

> There is no technology in itself, but in its relentless march it is accompanied by Man, without which it is nothing (Ellul 1954).

> Without being read, Descartes determines us, whether we like it or not (Jonas 2013).

> Indeed the world is ruled by little else. Practical Men, who believe themselves to be quite exempt from any intellectual influences, are usually the slaves of some defunct economist [...] It is ideas, not vested interests, which are dangerous for good or evil (Keynes 2016).

> To explode a myth is accordingly not to deny the facts but to reallocate them (McCown 2001).

The function of science is to produce proofs for propositions and to integrate these proofs into a kind of systematic theory or model the paradigm is then scientific. It is an evidence-based approach. "Technological paradigms" came into play when science-push models of innovation were replaced by demand-pull models, which justified a more international, market-driven economy. According to Casagrande-Seretti et al. (2019), a technological paradigm is a concept that asserts that improving technical performance alone is not enough to ensure the success of products and services. In addition, it should open up new and rewarding avenues of research for academic researchers, provided that the directions of study are clearly identified. However, Bayesian belief refers to a probabilistic estimate of an emerging phenomenon, and therefore does not take the unexpected into account. For the future, the processing of sensory information, which can be modified by new information inputs, leads to probabilistic analyses that enable decision-makers to make predictions that are largely limited to increments (business as usual).

Laperche (2020) recalls that Dosi (1982, 1988) proposed the concept of paradigm and technological trajectory for innovation. A technological paradigm is defined as "a set of procedures, the definition of relevant problems and the specific knowledge related to their solution" (Dosi 1982)[2] or "a model for solving selected technological problems, based on selected principles derived from the natural sciences and on selected material technologies"[3]. The paradigm is defined on the basis of the state of the knowledge base, available capabilities and specific economic and technological trade-offs. It envisages the realization of promises (as with the scientific paradigm). Dosi (1982) defines the technological trajectory as "a model of problem-solving activity. It is a set of generic technologies with a common dynamic"[4]:

> A technological paradigm is both an example (an artifact that needs to be developed and improved, such as an automobile, an integrated circuit, a lathe, each with its particular techno-economic characteristics) and a set of heuristics (for example: where do we go from here? In what direction should we direct our research? What kinds of knowledge should we encourage?)[5].

If we add caution, or even the fear of making a mistake, then the value of innovation depends on the conditions of the sector concerned, the institutional environment, the innovative efforts of competitors, the risk of obsolescence and the geographical proximity of the inventors. This value is also determined by

2 Author's translation from the French version of the text.

3 Author's translation from the French version of the text.

4 Author's translation from the French version of the text.

5 Author's translation from the French version of the text.

organizational characteristics, such as the company's absorption capacity and ability to combine innovations (mature knowledge can be fruitful, if it can be combined with more recent knowledge) (Capaldo et al. 2014).

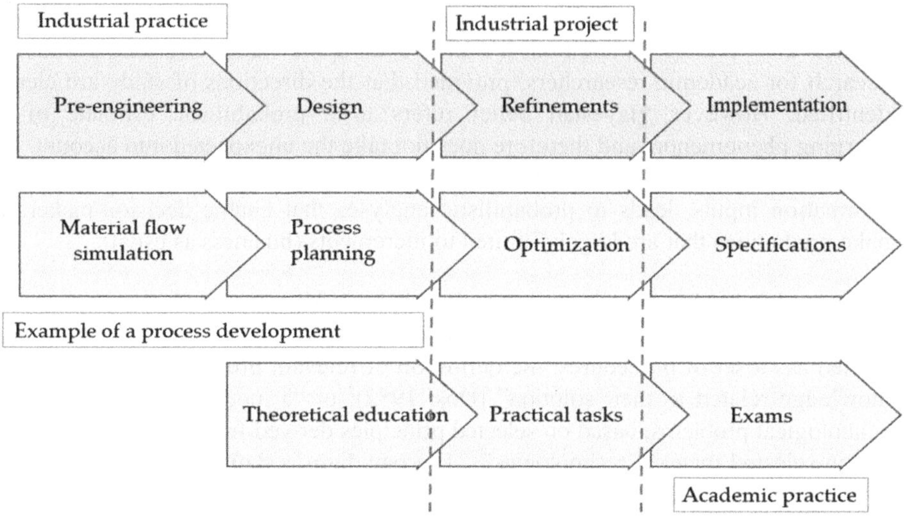

Figure 2.15. *Relationship between academic and industrial practices*

Thus, envisioning the emergence of a new technological paradigm involves several issues, ranging from strategic assessments and managerial actions to design decisions and technology-related choices. It requires covering the need for personnel adapted to changes, hence an obvious difficulty: training personnel in a disruptive technology that does not yet exist (or having the adaptability of the new technology to "old" knowledge or have agile and adaptive people)! The path to achieve this emergence involves both mastering the technology (see sections 2.1 and 2.2) and market uncertainties for companies. According to Norman and Verganti (2014), many companies encounter difficulties in maintaining their market leadership, because they are unable to interpret the potentialities of new technologies (with their relevance and their shortcomings) or to have suitable personnel. Figure 2.15 indicates the links between academic knowledge (including invention) and practices (Rentzos et al. 2014).

NOTE.– Technological paradigm and heuristics:

Such a paradigm constitutes the matrix of a series of technical solutions to preselected problems. It envisages both negative heuristics (trajectories to avoid) and positive heuristics (trajectories to follow). It is a research framework, applicable to different types of technology and different types of problems, indicating the direction of technical progress to be sought. "Within a paradigm, the design is constantly modified by incremental innovations, but these remain within the limits set by the paradigm. These innovations tend, in fact, to adapt the paradigm to local or contextual conditions" (Haudeville and Le Bas 2016).

2.4.3. *Innovation ecosystems*

> The legitimate use of contradiction consists, when two incompatible truths impose themselves on human intelligence, in recognizing them as such and making them the two arms of a pair of pliers, an instrument for indirectly entering into contact with the realm of transcendent truth, inaccessible to our intelligence (Weil 1955).

> In most cases, a theory considered new has been preceded by a long process of fragmentary attempts. Then comes a moment when the ideas that are 'in the air' are, so to speak, crystallized and find their full expression in the thought of a Higher Man who puts his personal stamp on them (Gaudin 2013).

> Provide us with candidates who won't know how to do it the first time, they'll want to get there all the more, we'll see if there's a need for training and what we can do about it together (Schwartz 2021).

Figure 2.16, taken from Granstrand and Holgersson (2020), reminds us that innovation means playing on a three-legged register with strong interactions: people, artifacts and their use.

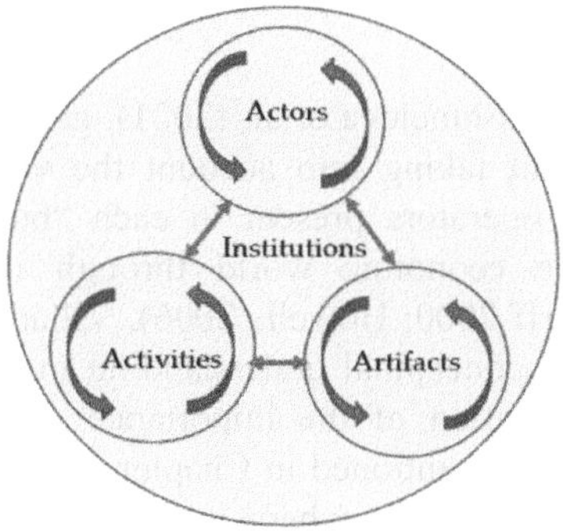

Figure 2.16. *Innovation ecosystem*

The number of startups and their survival has an impact on the economic wealth of states. Development criteria are important for this purpose (Volkmann et al. 2010). A startup's business idea represents only half of its future success, and the rest is associated with the quality of the organization and its performance discipline (Cohen and Feld 2019).

By avoiding basic mistakes and adopting rational rules, we can increase the survival rate of small innovative companies in the early stages of their development. But, as we explore the application, a survey (Kollmann et al. 2016) of the problems perceived by startups has been carried out (see Figure 2.17) and provides an overview of the challenges they face (survey participants indicated their top three challenges). Sales and customers are the most important, at 19.5%, followed by product development (17.1%) and growth (16.6%).

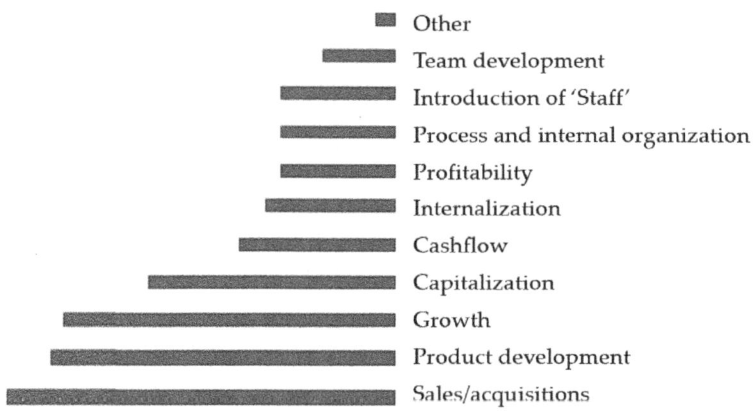

Figure 2.17. *Challenges facing European startups*

Figure 2.18, taken from Shmeleva et al. (2021), gives an overview of the entire innovation system (without taking into account the very large number of local, national or international operators present in each "bubble"). While universities foster relations with the economic world through intermediary organizations (Etzkowitz and Leydesdorff 2000; Howells 2006), Villani et al. (2017) demonstrate the existence of a lack of conceptual understanding in the intermediation process, linked to a weak consideration of the importance of cognitive proximity. This difficulty is the same as that mentioned in Chapter 1 of this volume, in section 1.3, in the research–SATT relationship, where it is necessary to find an interlocutor capable of overcoming this difficulty. Moreover, according to O'Kane (2018), researchers believe that transfer operator executives encounter difficulties in mastering the capacities linked in principle to their expertise in terms of market knowledge, business development and industry connectivity.

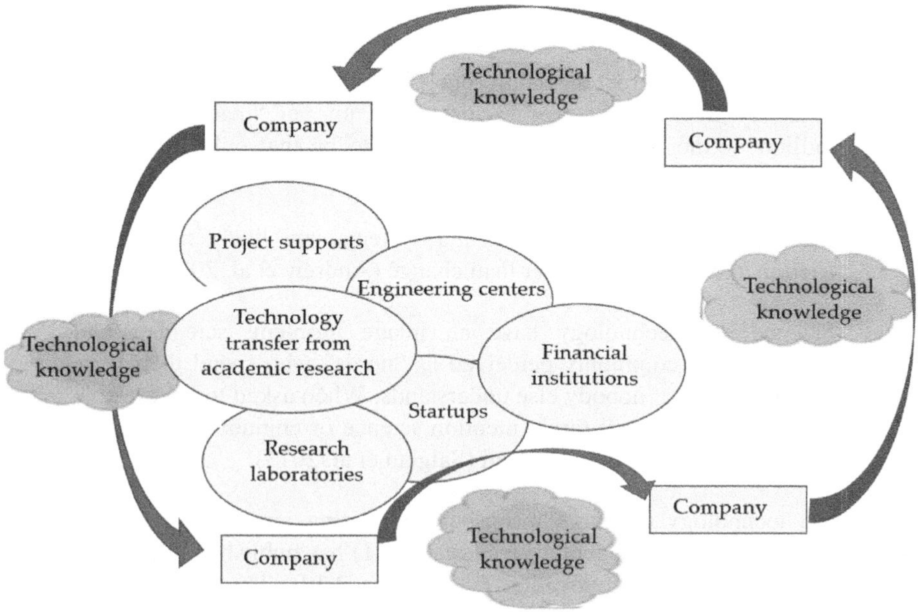

Figure 2.18. *Overview of the innovation system*

NOTE.– The role of universities:

In the ecosystem model, university technology transfer operators (Shmeleva et al. 2021) can do the following:

– initiate project ideas;

– give expert, analytical and prognostic analysis;

– build relationships between project participants;

– analyze information flows, construct a common information space structure;

– give project implementation and support;

– give performance management and tuning.

Transfer operators control most of the activities, resources and skills in the transfer process value chain, with the aim of maximizing revenues from the commercialization of academic results, and their role is, according to Matt and Schaeffer (2012), to manage a linear, unidirectional process. Today, these operators are forging alliances with the aim of disseminating research results more widely and rapidly. The technology transfer process is more interactive and multidirectional.

2.4.4. *Innovation accelerators*

> In the area of entrepreneurship, one can often hear open disdain for scholarly work because professors have not yet "made their first million" – the foremost argument seems to be that only experience counts (Frese et al. 2012).
>
> Incentives, without knowledge and systems, are likely to produce friction and frustration rather than change (Andrew et al. 2021)[6].
>
> Science and technology have an image problem: scientists and engineers are commonly perceived as "nerds" who spend their lives doing things that nobody else understands. When asked to list creative activities, people will rarely mention science or engineering, offering arts and humanities areas instead (Salgian et al. 2013).

Digital technology can be a means of stimulating innovation. Innovation has become more open and collaborative. Nesta (2021) has published in its latest report 13 methods to stimulate innovation (see Figure 2.19). These are the following methods:

– startup accelerators, where we invest (at risk) over a short period (between 3 and 12 months) in well-targeted young companies;

– early settlement is an emerging, proactive, iterative and market-driven approach to overcoming emerging challenges;

– challenge prizes: through an open competition, well-defined challenge prizes aim to mobilize and engage the community of innovators in solving a particular problem or challenge;

– participatory through numerous small contributions from a large group of people;

– experimentation is a means of testing an innovation with a group of its target customers;

– futuristic methods include tools that aim to predict the future in order to better prepare for it;

– impact investments in organizations that develop products and services or use their operational infrastructure to make a positive contribution to society or the environment;

6 Author's translation from the French version of the text.

– innovation mapping which informs decision-makers about innovation zones in their region;

– people-oriented results: the 100-day challenge is a structured innovation process based on working groups that include front-line users of public services;

– prototyping is a means of developing, testing and improving ideas at an early stage;

– innovation laboratories dedicated to structuring and integrating methods and practices in public authorities to solve social and public problems;

– subsidies for social innovations to accelerate the scaling of services offered by nonprofit companies;

– standards provide a framework for choosing the right approach to understand whether an innovation works. They provide a structure for determining whether the innovation is making a positive difference to society.

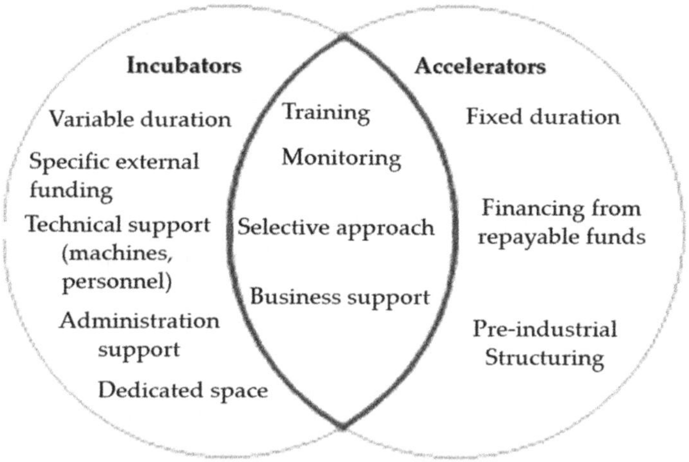

Figure 2.19. *Innovation accelerators*

2.5. Innovation models

> Scientists thus become even more determined promoters of somnambulistic progress (Stengers 2014).

> In the event of the loss of skilled labor resources, "it would take at least a whole generation to repair this misfortune" (Saint-Simon 2012).

> The structure of power is interests and policies, the activities of institutions, parties, pressure groups and bureaucracy. It requires a look at the nation as a whole. The role played by the political system to a large extent determined by the nation's economic and social structure, and by the values of the main social groups. Conversely, society is shaped by political decisions and processes (Hoffman 1963)[7].

Western living standards depend to a large extent on the ability of companies to create globally competitive products and services. Innovation is and will continue to be an important key to solving productivity problems. As long as we have been able to exploit the model (and as is largely the case today), societal or industrial challenges have been addressed through links between production and public economic development policies (Lesourne and Randet 2016). However, according to Ruano-Borbalan (2017a, 2017b), scientific knowledge production remains for the most part disjointed from strict economic "utility". Indeed, in order to address major trends, research work is engaged in long-term "programmable" activities.

NOTE.– Linear models:

The technology push model sees innovation as the result of industrial policies; it supports key technology policies (e.g. nanotechnology, but with mediocre impact, see Frochot et al. (2020)). With Schumpeter (1983), this bottom-up model assumes that innovation stems from science and technology. According to Badillo (2013), Schumpeter proposes two versions: one in which it is the company that is the actor of innovation and the other in which it comes from the R&D departments of large companies. On this basis, Dosi (1982) developed the concepts of technological paradigm, and Freeman et al. (1982) developed technico-economic paradigm. In 1966, Schmookler proposed a demand pull or market pull model of innovation. While linear in form, it starts from the consumer's demand (see Tucsnak and Weiss 2014). According to Badillo (2013), these two principles of action led to the interactive model of innovation (see Rothwell and Zegveld 1982) associating technology and joint user demand. "The overall innovation process can be thought of as a complex set of communication paths over which knowledge is transferred" (Trott 2008).

This raises the question of an invention's desirability. Steenkamp (2002) shows that there are several categories of public adopters of a new invention, and that these depend on the national culture of each country (see Table 2.1). In some countries, such as the UK, this situation makes it possible to take risks relative to other more timid nations, such as Spain.

---

7 Author's translation from the French version of the text.

Country	'Adopters' (%)	Convinced followers (%)	Trained followers (%)	Latecomers (%)
USA	16	34	34	16
UK	23.8	43.4	26.4	6.4
France	15.1	25.5	35.6	23.8
Germany	16.8	26.1	34.2	22.9
Spain	8.9	34.1	43.9	13.1
Italy	13.4	30.8	41	14.8

Table 2.1. *Distribution by adopter category. For a color version of this table, see www.iste.co.uk/andre1/knowledge.zip*

While incremental innovation brings an improvement to an existing product or process, a DI involves launching a radically new product or process. When the innovation truly overturns the rules of the market, we speak of DI:

> At the crossroads of management science, project management and knowledge management, the innovation process is seen as a project and a space for the creation and capitalization of a body of knowledge, based above all on methodological approaches and organizational arrangements that encourage interaction and creativity (Wannenmacher 2020).

A co-design approach stemming from the technology phase (see Chapter 1 of this volume) must enable the integration of the needs of end-users into the innovation. This alone cannot guarantee wider appropriation, whether in terms of number of users or technology transfer. "Once conceived and used, the object continues to evolve as it is constructed and reconstructed in its socio-technical network" (Akrich 1997)[8].

However, some may invest too early, fail to follow fashion trends and fail to take into account the economic maturity of an emerging technology. For Alti-Patrimoine (2022), there are several reasons why the price of a promising theme may stagnate or fall:

8 Author's translation from the French version of the text.

– companies in this technology are not yet mature, not yet sufficiently structured, their net income and sales may still be too low or negative, or they may even be in the process of being created;

– companies may be considered to be mature, but the market has yet to recognize their potential and value, because investors are not yet interested.

For example, the concept of indoor farms is attractive because it enables fruit and vegetables to be produced anytime, anywhere, with high, stable yields, but for the moment (and probably for a long time to come), the cost of production is too high.

2.5.1. *Open innovation*

> Schumpeter's cumulative innovation is the primary source of growth: any individual will tend to underinvest in innovation, because he does not internalize the improvement in knowledge that his innovation brings to society and the fact that other innovators in the future will be able to build on this knowledge (Aghion et al. 2020).

> The world has always been full of people like Piketty, who feel entitled to dictate how others should live, so that their concepts of social perfection can come true. If one day we can break the bad habit of paying attention to them, we'll be much better off (Leef 2022).

> The economics of creativity and the economics of problem-solving are two very different fields. While creativity is by its very nature not functionally located within a company, and can be the work of individual researchers, outsiders or displaced people, users and problem-solving processes are more often organized and collective, in places dedicated to this function (Foray 2002).

In today's economy ecosystem, success is not just personal, but involves participating with other organizations in the innovation process because it makes it easier to innovate for ourselves. This collaborative, partnership or participatory approach (Vinatier and Morrissette 2015) gives rise to the figure of the "practitioner-researcher", who envisages producing knowledge not only "about", but also "with" practitioners, and guiding practice via the knowledge of action mobilized (Desgagné and Bednarz 2005). So, everyone can be a beneficiary and a contributor in a positive "game".

As a result, enhanced innovation, in a field where the principles of collective intelligence must come into play, is considered possible despite scarce economic and material resources. The underlying idea is that the "sum" of individual creativities, intelligently arranged and exploited, can become the basis of an "immense" collective intelligence and a lever of innovation and performance for society. Nevertheless, this remains a challenge that relies on open and frank cooperation, which is a guarantee of sustainability, flexibility and performance:

> Human beings, faced with common adaptive problems can decide, from mental schemas, to adapt or not. These mental patterns make possible a permanent oscillation between an assimilation of the real on oneself and an accommodation of this self to the real, and aim at a regulation of behavior by the equilibration of regulating adaptive antagonisms. The adaptive capacity allows human beings to oscillate between the binary, even ternary clusters of adaptive antagonisms such as identity versus otherness, openness versus closure, centration versus decentration, proximity versus distance (Morace and Coadour 2018).

The appeal of open science for Gold and Morgan (2019) lies in the fact that it combines incentives for strong academic competition with those of the business world and enables partners to assume risks and rapidly transfer results from the research sector to the commercial sector. In this way, open innovation, in cooperation with different partners and trusted customers, enables us to better understand and anticipate market needs. In today's context of volatile demand, fierce competition, rapid technological change and scarcity of resources and talent, it is all the more advantageous for companies to use this approach. For companies, cooperation with competitors is not easy, but there are platforms that can bring all stakeholders together in a digital workspace that enhances everyone's productivity.

In these interactions, Waikar (2022) shows that teams succeed when they balance creativity and concentration. Goldberg (2002) reminds us that "in an organization, there are tensions between people who have incongruent ways of thinking about a particular problem to be solved". He goes on to write that "intellectual diversity is good for creating novelty and solving problems creatively, but not necessarily for effective coordination". He showed that the most successful teams modulated their cognitive diversity to suit the task in hand. Thus, groups whose ideas become divergent for ideation, but more convergent for coordination, are those most successful in delivering their projects on time and to the customer's satisfaction.

Moreover, according to Heracleous and Robson (2021), "rebellion" is essential to creativity. Thus, organizations have rules and policies designed to promote stability, predictability, efficiency and productivity (including quality). Group members who fail to respect these constraints tend to be seen as rebels. Failure to take into account the role of these people, with their divergent behaviors, can deprive organizations of a powerful source of agility, insight and innovation.

2.5.2. *Knowledge management*

> Irony has a few advantages. It brings out the essence of a question. It reveals how the question concerns us. Used after tedious argumentation, it underlines the possible significance of the results obtained and makes even the most carefree readers aware of their novelty (D'Espagnat 1983)[9].

> Conducting a rigorous, respectful and socially relevant participant observation. Completing and submitting reports, requests for funding and for funding extensions for deadlines that are always multiplying and getting closer. Analyzing data, co-constructing results and guiding technical choices whose realization is carried out at the pace of independent production chains. These are just a few of the paradoxical tasks – at least in light of their respective temporalities – that ethnologists adhering to the anthropotechnological approach must navigate when they are working in the context of the development industry (Geslin 2017).

> The reality is quite different. Every invention allows us to solve old problems, but new ones immediately appear, in a different field. We are caught up in a movement that, far from slowing down, is accelerating. Let's not expect us to be able to make a few efforts and then rest. Our very progress requires us to seek further (Berger 1964).

Its aim is to transform data into usable information that will serve as the basis of the company's knowledge base. It focuses on the links between individual and collective knowledge, and coordinates the resulting actions:

> Knowledge is in fact a set of structured information, oriented on a subject and validated by rules or experience. So-called "explicit"

9 Author's translation from the French version of the text.

knowledge, which can be formalized and therefore computerized, represents only a small part of the total body of global knowledge (Fernandez 2001).

KM involves exploring and combining the information produced internally with that resulting from a watch on the themes concerned by the invention, which contributes to its technological, economic, sociocultural or even legal environment. These different processes of collecting, processing and analyzing the existing situation therefore have the mission of generating actions that benefit the company. The challenge is to know how to provoke and support long-term innovation initiatives in a learning organization context (idea of collective intelligence).

Wannenmacher (2020) provides a critical analysis of KM definitions in Table 2.2.

Definition type	Definition	Critical analysis
Utilitarian	Processing the existing information to respond to an actual need	Confusion between information and knowledge More like economic intelligence
Functional	Managing the knowledge life cycle following the emergence of an idea: formalization, validation, diffusion, reuse, valorization, etc.	Most widespread definition. Limited vision because when a human being creates knowledge, they do not cut up the processes
Operational	Combining knowledge and know-how in the process, products, organizations, in order to create value	Pertinent definition if we consider that skills and knowledge produce value when they are combined with other factors
Economic	Valorize the firm's intellectual capital, which is considered an immaterial asset	The question of the analysis of immaterial value is tricky because knowledge escapes classical rules of economy

Table 2.2. *Critical analysis of KM definitions*

In his paper, he also makes a number of recommendations to follow up on his analysis of the failures of KM approaches (see Table 2.3).

Cause of failure	Recommandations
Approach too oriented towards the technological tool, to the detriment of human and social components	First create exchange communities and flows around strategic activities, then install a tool to support the process
Too complex an approach that favors formal and mathematical modeling too much	Favor simple, participatory approaches adapted to the specific context
Exotic approach not adapted to the local culture	Work on interculturality, understand behavioral differences, take into account the company's "house" culture, promote decompartmentalization
Approach wanting to codify everything in knowledge bases	Accept the tacit and non-codifiable character of an important part of the experience. Work on the mapping of crucial expertise and adopt several transfer mechanisms: socialization, tutoring, narration…
Approach stemming from the solitary work of an analyst	Favor participatory, incremental and iterative approaches, even if it means being less ambitious at the start but more mobilizing
Approach by costs/expenses, experienced as an obligatory chore	Integrate the capitalization approach into operational processes. Linking KM to strategy and economic objectives
KM approach as an object of conflict between major functional departments	Promote a systemic and multidisciplinary approach. Obtain strong and official support from general management and create a steering committee bringing together members of the various departments impacted
Approach hindered by an authoritarian management mode and culture focused on power	Think of the project as a change management process. Show the first tangible benefits quickly, to facilitate deployment later
Knowledge management approach displayed which is in fact reduced to documentary archiving	In the upstream phases of the project, carry out a discussion and an update on the meaning of the main concepts used and on their scope

Table 2.3. *Analysis of KM approach failures and proposed recommendations*

It should be remembered that the presentation of constraints, phases and project players can lead to misunderstandings, particularly at the start of a project, due to the different cultures and knowledge possessed by the partners involved in the operation (see Chapter 1, Volume 1). However, according to McKinsey (2022), these considerations should be extended to the use of digital technology. Thus, seven elements would define this "new" data-driven enterprise:

– data are integrated into every decision, interaction and process;

– data are processed and delivered in real time;

– flexible data repositories will enable data integration and promote employment;

– the operating model will treat data as a product;

– the role of the data manager will be expanded to generate value;

– data ecosystem memberships will be the norm;

– data management will be prioritized and automated for confidentiality, security and resilience.

Deloitte (2021), for example, highlighted the rise of smart factories because of the rapid convergence of transformational technologies. It would seem that the connected and flexible factory that is activated by a constant flow of digital data from operations is an unavoidable future of manufacturing. To innovate, companies need to cooperate extensively, including with suppliers, customers and universities. This level of coordination takes time and requires leadership. Agencies such as DARPA (Defense Advanced Research Projects Agency with a budget of US $3.5 billion/year) bring together established companies with start-ups and industry experts. They facilitate the sharing of knowledge, provide third-party validation of new technological directions and support the development of interaction platforms.

2.5.3. *Research versus industry*

> Technology, in addition to its undeniable progress, does it serve to fuel dreams and make applications more acceptable to the majority, or does it give a few science-fiction fans to influence research programs? (Cornu 2008).

> It's out of fear of the unforeseen and its disorientating power that we reinforce methodological protocols. But the more we reinforce them, the more paralyzing the occurrence of the unexpected becomes (Lavis 2021).

> Life is an experimental journey, carried out involuntarily. It is a journey of the spirit through matter and, since it is our spirit that travels, it is in it that we live (Pessoa 1992).

The transfer of technology from public sector research to industry is linked to various conceptualizations of innovation processes. The so-called model of innovation, reflected in Vannevar Bush's (1945) report *Science: The Endless Frontier*, sees public research as serving industrial development in a unidirectional way, without the proactive participation of business. This simplistic view has been replaced by various interactive models of innovation.

> These models also recognize that technology transfer is a two-way street in which researchers and companies interact intensively and draw quite frequently on basic research external to the company (Palmberg et al. 2010).

A high proportion of industrial innovations in high-tech sectors are directly based on recent academic research, even if the invention itself does not originate from this world. By seeking to meet the perceived needs of various production communities, new methods of conceptualization and measurement of short-, medium- and long-term impacts can be used to assess the success or failure of socio-scientific and interdisciplinary research (Rau et al. 2018). Innovation processes are, as we all know, multiple in terms of subject, nature, etc., and are often messy, nonlinear and involve the cooperation between multiple people (Van de Ven et al. 1999). They benefit from the integration of diverse knowledge, resources and tasks across boundaries (Love and Roper 2009).

Research and development (R&D) must be protected, developed and passed on to industry and society. They are steered by a cross-functional team with two specialized roles: academics and people in charge of transfer. Between these two (sometimes obliged) partners, there are differences in power and boundaries that may be mental (Zerubavel 1991) and competences. According to O'Kane (2018), boundaries that are too weakly permeable can compromise knowledge sharing and the production and creation of value associated with innovation. At the same time, Clarysse et al. (2011) find that relay activities play only a minor indirect role in relation to scientists' entrepreneurial capacity.

Bozeman et al. (2015) propose a transfer model (see Figure 2.20) based on a concept of contingent efficiency of technology transfer. They suggest, along with Anderson et al. (2007) and Kraujalienė (2019), a number of recommendations addressing general issues relating to the evaluation of the effectiveness of technology transfer:

– making the most of the model presented;

– identifying expected impact limits with an understanding of the influence of the intervention via the transfer agency;

– further developing indicators of scientific and technical human capital;

– correlating process reforms and activity measurements;

– distinguishing the relative importance of indicators associated with transfer by universities, government laboratories and industry;

– increasing attention and development of efficiency criteria of public value.

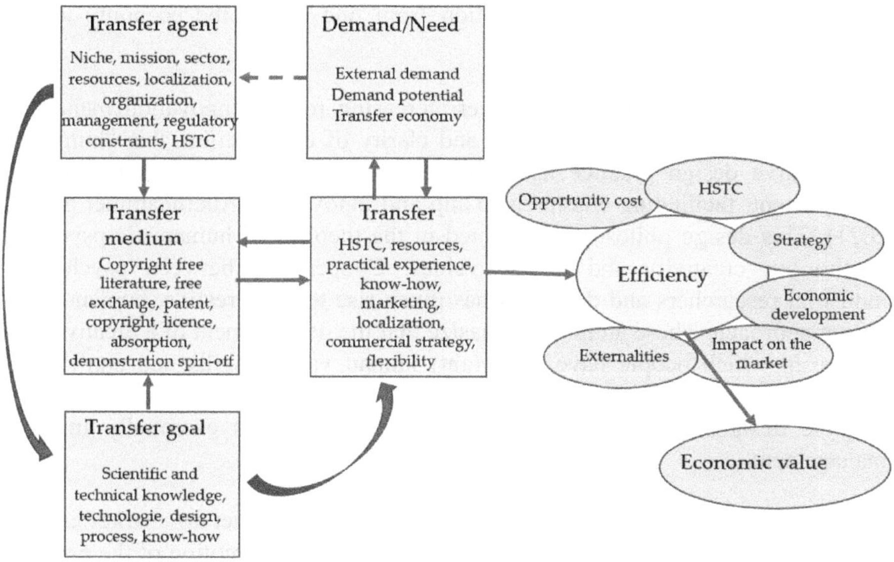

Figure 2.20. *Transfer model based on a concept of contingent efficiency (HSTC: human scientific and technical capital)*

2.5.4. *Design-thinking*

Design uses culture and art to transform a technology into a useful object that will generate a commercial transaction. The use of design-thinking in an organization requires that those involved, and those responsible for innovation to understand the history of design, and to be aware of the different possible approaches to the discipline of design. It would therefore be astute to combine all these forms of scientific expertise of design-thinking as a pragmatic, concrete method, and to show the extent to which design-thinking also

> contributes to the conceptual renewal of management and even organizations (Szostak 2021).

> The formulation of a problem is often more essential than its solution, which may be merely a matter of mathematical or experimental skill. To raise new questions, new possibilities, to regard old problems from a new angle, requires creative imagination (Einstein and Infeld 1938).

> It's not the rheumatism of old age that we're suffering from, no, but the growing pains caused by changes that are far too rapid a pace, disorders linked to the transition from one era of the economy to another (Keynes 2021).

Design-thinking is playing an ever-increasing role in innovation management discourse. From its theoretical roots and clarity of construction, this "humanistic" and creative design practice has evolved into a holistic approach to innovation management, facilitating entrepreneurship and innovation (Auernhammer and Roth 2021). This design philosophy is rooted in the theories of humanistic psychology, notably on creativity and human values. Cooperation between psychologists, industrial researchers and designers has given rise to this creative, human-centered design approach. These authors emphasize that the development of a creative design culture in which people have the (right) human values, skills and confidence to collaboratively identify ongoing emerging problems and needs, and contribute to tangible designs, generates an era of innovation and is essentially innovation management.

Using the latest technology or bringing a new product to market is not as important to success and profitability as the design and promotion of the experience that accompanies the proposed innovation (Léac 2022). This experience tends to reinvent marketing around the promotion of a particular, unique and new experience. Design-thinking with its promise of designing experiences seems to be one way in which companies can respond to these changes (see Figure 2.21).

Design-thinking can be extracted and separated from traditional design practice. Today, a widely accepted nomenclature can describe the design process, with its positive effects on communication between designers and managers:

> However, due to a widespread implicit cognitivism in the literature, scholars find it difficult to explain the cultural and experiential qualities of design thinking and it tends to be presented as a fundamentally cognitive, problem-solving activity (Rylander et al. 2021).

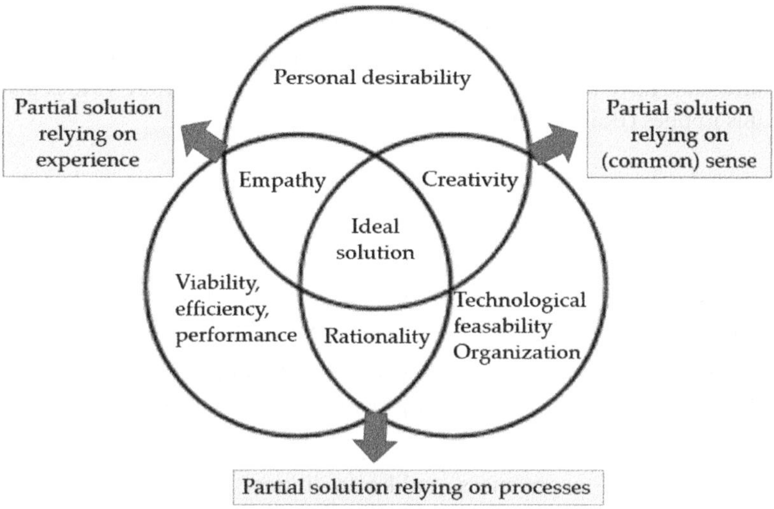

Figure 2.21. *Design-thinking*

These authors consider that these cognitivist tendencies blur the appropriate focus on designers' creative practice, its theorization in the form of a theory of practice. However, it is a process of meaning that emphasizes imagination and improvization in its core activities and explains how sensibility is developed and nurtured.

Micheli et al. (2021) and Johansson-Sköldberg et al. (2013) show that design-thinking has been described by managers as one of the best ways to be creative and to innovate. For designers, this process is taken for granted with several modes of thought and/or practice that have distinct epistemological roots. For managers, design-thinking would rarely be referred to, which would hinder the construction of cumulative knowledge. From a practical point of view, the stages of a design-thinking approach are:

– identifying a problem and understanding its environment;

– finding the concept, the idea that will solve it;

– designing the form that will embody this concept.

NOTE.– Epistemic virtues:

> Intellectual honesty, integrity, the ability to subject what we put forward to the criticism of others, a certain and indispensable mistrust of our senses and memory, the ability to consider alternative hypotheses, the

> practice of constructive doubt, recognition of the fallible nature of our knowledge, and so on. For it's not enough to possess knowledge: we must also have the determination to use it, and the epistemic virtues that this implies (Baillargeon 2012).

In concrete terms, working in design-thinking mode is organized around three logics:

– a logic of co-creation: a company that puts design-thinking at the heart of its business does not make its departments work in isolation, but instead establishes a "cross-departmental" logic that encourages collective intelligence;

– intellectual gymnastics, alternating phases of intuition and analysis, with an open/close approach (as in any creative process such as brainstorming);

– a major emphasis on field research (ethnographic observation), which offers a full understanding of experiences, unlike traditional quantitative and qualitative studies.

Put simply, creating customer experiences is financially profitable. Léac (2022) reports that a 5% increase in customer loyalty can translate into a 25–50% increase in profits for the company:

> By creating unconditional fans among customers, a company builds a real marketing team who talk about their experience to their peers – far more effective than any advertising!

Even if the invention may be far removed from this operation, design-thinking is developed in several stages as follows:

– defining the problem to be solved, prioritizing the project and determining what will ensure its success;

– researching: reviewing the history of problems encountered, collecting examples of failures, identifying supporters, investors and critics of the project, talking to the end customer;

– identifying end customers' needs and motivations, generating as many ideas as possible to meet these needs without judging them, brainstorming;

– prototyping: combining, cross-referencing and refining ideas, creating drafts/models/prototypes, receiving feedback from potential and existing customers;

– selecting: reviewing objectives, transferring ownership of the selected idea to the person who came up with it, choosing the most surprising, new and economical idea;

– implementing: drawing up the action plan and assigning responsibilities, determining the necessary resources, delivering them to the customer;

– learning: receiving feedback from the end customer, determining whether the validated solution meets the initial objective, identifying sources of improvement.

The NAE (2022) reminds us that the search for questions, like serendipity, cannot be programmed. We therefore need to move from funding projects to people, "not just as a mantra, but as a fundamental element of effective techno-scientific research".

2.5.5. *Fab-Labs*

> To what light do we first recognize the value of these sudden syntheses? To an unspeakable clarity that puts security and happiness into our reason (Bachelard 1963).

> My scientific work had never been oriented towards solving theoretical problems [...] at least towards contributing to the resolution of societal problems by means of science, including, naturally, the development of theories on the subject. This is an inversion of the priorities of the scientific community (Wisner 1995).

> FabLabs introduce new ways to approach innovation, fabrication and the distribution of objects around us. Some commentators of the movement consider them to be key players in a new industrial revolution. Without taking a position with regard to this statement, it is certain that, because of the philosophical movement underlying it, the international network of FabLabs is opening up new possibilities. New types of products are being created and often these products would never have been able to be "born" without an alternative approach to design and "industrialization" (Geslin 2017).

Launched by MIT's Media Lab in the late 1990s (Gershenfeld 2012; Fablab 2016), Fab-Labs are places open to the public where all kinds of tools are made available for designing and making objects. For Schneider and Lösch (2019), Fab-Labs can, in a very positive view, be analyzed as an empirical case of democratization of the creation of the future. Their charter (GoTronic 2016) is recognized worldwide via MIT. It is well known that it is minorities who innovate. These active minorities present in Open-Labs or Fab-Labs (Fabrication Laboratory) can be creative powers capable of setting in motion a delocalized process of influence aimed at the silent majority. With low-cost machines, experimentation can expand application markets, while leading in turn to social adaptations (Mavri 2015).

At a time when we are going through a major economic and social crisis with its environmental problems, a global pandemic, where companies are called upon to produce more and "better", "cleaner", and more frugal ways, with fewer resources in production sites where hyper-connectivity takes precedence over conversation and friendly exchange, if possible in-depth, some organizations wonder about a reinforcement, or even a return of inventiveness, creativity, generosity and spontaneity in relationships between humans (Gaines 2019). "The Internet is bringing collaboration back into fashion at the very moment when the ideology of all-out competition is revealing itself for what it has always been: a truncated and limited vision of society" (de Grave 2015). It stems from this positioning, associated with a "2.0 mentality", a recent research where society is supposed to reflect more "natural" tendencies and not a set of obstacles and rules imposed on citizens.

Access to open source technologies and the development of information and exchange networks via the Internet have created another form of freedom, linked to making artifacts at home, or in a Fab-Lab. The change in culture associated with this phenomenon is seen by some as a form of organizational revolution in 21st-century society, because it is likely to change manufacturing methods by relocating them closer to the user. For Botsman (2013), the collaborative economy, into which the *maker* movement can fit into, is defined as "networks of connected individuals and communities, as opposed to centralized institutions, that transform the way we produce, consume, finance and learn" (see also Petropoulos (2017)). It encompasses, according to Terrasse et al. (2016):

– collaborative consumption, that is:

- the organization exchanges and consumption (group purchasing, private rentals, resale and donation of objects, bartering, local exchange systems, currencies, etc.),

- housing (shared housing, self-build, home exchanges, etc.),

- energy (cooperation in energy purchasing, etc.),

- mobility (car-sharing, car-pooling, etc.),

- food and agriculture (food self-production, short circuits, consumer cooperatives, etc.);

– repair and manufacture of objects (hackerspaces, Fab-Labs, Tec Shops, Repair cafés, etc.);

– project (crowdfunding, citizen investment, solidarity finance, etc.);

– education and knowledge (digital content sharing, Massive Open Online Courses (MOOCs), Wikipedia, etc.).

The promoters of the collaborative economy draw on several schools of thought (Terrasse et al. 2016):

– The "free" movement is characterized by a shift from a principle of ownership (property rights and copyright), to one of free, unrestricted and universal access for users to use their knowledge, skills and material resources via open systems; this is the domain of makers.

– The functionality is an economy that "aims to optimize the use of goods and services. The economic objective is to create the highest possible use value for the longest possible time, while consuming the least possible material and energy resources" (Stahel 1986).

– The gift economy covers modes of exchange in which the goods and services in circulation are not considered for their value.

This new context is part of what Servier (1991) considers utopia, a dream to be reintegrated into reality, because of certain specific characteristics:

– utopia is associated with a dream, an idea of the future, a sense of originality;

– the feeling of belonging to a community of thought, partially isolated and tolerant of others;

– forms of desire for a return to a certain "purity".

Unlike older forms of utopia, this new context is different in that it does not seek isolation or nostalgia for the past. In this sense, they are open, ephemeral organizations, with no specific desire for idealization, or even "normalization" or the use of identity symbols, with whom there is scope for progress through exchanges at all levels (Bina et al. 2020).

But at the same time, according to Moussion (2015), in a global democratization, the question of intellectual property (in the image of pirated copies in music) arises. He writes:

> Indeed, the legislator has set limits on the exercise of intellectual property rights. These limits are listed exhaustively in the Intellectual Property Code. With regard to copyright, article L.122-5 sets out a list of situations in which the author may not prohibit representations, copies or reproductions of his work. The exception for private copying set out in no.2 of this article is retained only if three conditions are cumulatively met. The copy:
>
> – is made from a lawful source, that is, from an original work legally acquired by the copier;

– for strictly private use only;

– does not conflict with a normal exploitation of the work and does not prejudice the legitimate interests of the right holder.

Lhoste (2017) reminds us that the notion of compromise already exists in the process of institutionalizing Fab-Labs. As Lhoste puts it:

> The most passionate have to make compromises to enlist new players. They broaden the range of activities on offer, with the aim of getting "everyone in the door" [...]. The everyday objects on display invite the curious to come and learn how to use the machines, while the codes and conventions reflect the kind of socializing typical of makers' communities.

For Pépin (2019), when making objects, the focus is on the skills and resources that Fab-Lab members can mobilize. Each person "must perceive themselves as capable of creating artifacts with their own skills, or of acquiring the skills they need to do so". If necessary, knowledge sharing can enable "learners to refine their techniques and improve their skills".

2.6. Inventing/innovating

> In a civilization saturated by technology, there is no longer any adventure; all that remains is the risk of being late (Sloterdijk 2006).

> We don't have to know where to start. We don't know what makes up our world, who acts in it, or what hardships they put each other through. Nor do we know what is important and what is insignificant, and what causes the shifts we observe around us (Latour 1984).

> In the experimental sciences, to explain is most often to "unfold", i.e. to create unprecedented situations in which what is being questioned demonstrates what it is sensitive to by deploying behavior that we didn't know it was capable of *a priori* (Stengers 2014).

In this field, it is often necessary to bring together knowledge from different methods and cultures. Interdisciplinarity reigns supreme. This is where the concept of convergence continues to unfold (see Figure 2.16(a) and (b)). This means we have to deal with the heterogeneity of the knowledge and data we call upon by moving away from reductionism. The question is to take everything into account, while imposing constraints (it is the engineer's job to try to optimize time and means, while seeking the minimum loss or, at the very least, deformation). Nazin (2016) proposes that "with a metalanguage, it is possible to envisage the integration of various theoretical sets, but above all it becomes possible to develop objective means, to compare different positions about a given problem and, by the same token, to make its application safer".

But will we have the time, the motivation (as in the case of biophysics) to transform a local interdisciplinarity into a new discipline, with its own paradigms? Will this new discipline be able to manage complex systems? In any case, the objective must remain as proof, understood as an established connection of precisely formulated, objectively existing ideas, avoiding lateralization and misappropriation. This connection of ideas belongs to the realm of the possible (Anisov 2020). This author reminds us that proof is not necessarily convincing for everyone, but "facts are stubborn"!

According to Rothwell (1992, 1994; Malecki 2010), there is a transition between the different "generations" of R&D which is now well known, with a strong evolution in innovation modes:

– technological push following the linear model (first generation);

– market pull or needs pull, with R&D based on customer needs according to customer needs (second generation);

– interaction or coupling between science, technology and the market (third generation);

– functional integration and overlap (fourth generation);

– system integration and networking (fifth generation).

2.6.1. *General framework*

> These skills are as follows: flexibility, adaptability, technological literacy, risk-taking, business thinking, and abilities connected with information management, cybersecurity, quality control and sustainability (Poszytek 2021).
>
> One of the driving forces behind successful innovation is technology transfer, in which the private sector adapts federal research for use in the marketplace (Link and Scott 2020)[10].
>
> Coordination of product and process changes requires the combination of specialized and tacit knowledge across functional boundaries. It's a process where accumulated experience counts (Malecki 2010)[11].

10 Author's translation from the French version of the text.

11 Author's translation from the French version of the text.

According to the classic vision of creativity founded by Guilford (1950, 1959) who defined it as "the complexity and the ability to generate new ideas" (Guilford quoted by Capron Puozzo (2016)) on the dichotomous principle divergence/convergence, the creative process begins with the recognition or creation of a problem. From there, a process of divergence begins, and finally ends, by convergence, in a new solution of the problem (when it exists). Figure 2.22(a), taken from Mathias (2019), presents a fast design method called double diamond, which uses these modes of divergence and convergence between input and output.

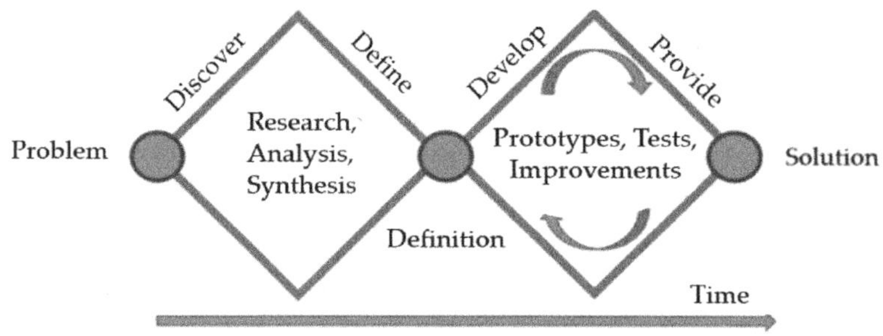

a)

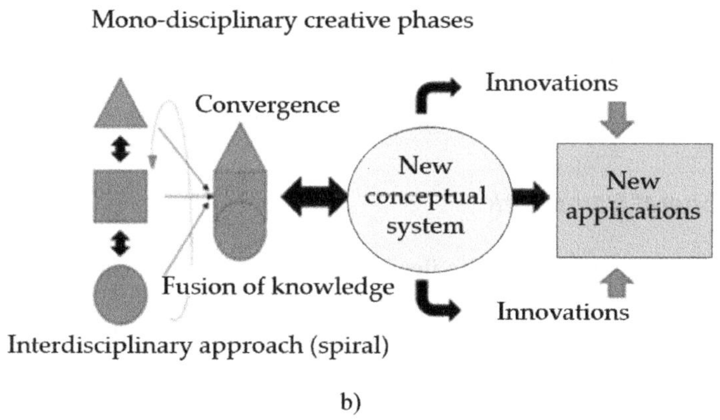

b)

Figure 2.22. *(a) "Problem-solution" relationship in innovation; (b) reminder on the principles of convergence/divergence (source: Buttin and Saffré 2016)*

The four phases of this design process provide intent and direction to the activities described. They can be applied by any designer:

– discovery (divergent): the task is clarified in this phase by exploring the problem space and identifying all possible design requirements;

– definition (convergent): a design specification is synthesized from the establishment of essential problems and requirements into a list of performance-based specifications;

– (divergent) development: possible solutions to the performance specification are evaluated, iterated and combined to develop a more complete design;

– deliverables (convergent): the chosen design is consolidated into a final specification detailing operation and manufacture, to be realized or communicated to others.

NOTE.– Transitions:

The anthropologist uses the term "mutation" to describe the current vagueness, the blurring of values, the indefiniteness of landmarks that are being diluted and the time in which traditional forms coexist, without new ones being legitimized. "It is a suspended time in which a transformation of the structures of industrial civilization takes place which makes the future uncertain" (Le Pogam 1997). Technoscientific modernity based on a triumphant rationality cannot ignore the emergence of these new disordered creativities. According to Chaurant and Brauer (2008), deviation from tradition is also a way of marking a person's originality, their uniqueness, of standing out from others:

> By not respecting the norms of the group, the individual does not act as a true member of the group and manifests his relative independence from his group (or from society). However, originality is in most cases a valued characteristic (we are more interested in a person who has 'something extra' than in interchangeable people). Paradoxically, adopting deviant behavior can then lead to an appreciation of the individual (Chaurant and Brauer 2008).

In fact, in the maker context, behaviors are evaluated rather positively, and deviance from "the system" is associated with creativity and its material consequences (see, for example, Fombelle et al. 2021).

Thus, with Bernard Stiegler (2014), we can speak with the makers of the (modest) emergence of a new industrial revolution, which would make it possible to envisage new, highly localized and decentralized forms of production, and revaluing know-how after Fordian proletarianization. Among the concepts retained, that of ideas which become "bigger" (Anderson 2012): the projects by expanding can become those of a group of "activists" who, by diffusion, expand them more and more to society. Apparently, in contrast to such a view, the integration of digital technologies suggests that automation will become widespread and wage employment will decline, if not disappear altogether (Staune 2015). For Giget (2013):

> We are currently experiencing a paradox. We're living through a period of considerable technological advance: we've surpassed 12 million researchers worldwide, two million patents filed last year and 110,000 scientific journals. But at the same time, progress in new technologies are having a redoubtable impact on jobs, destroying them on a massive scale.

In fact, following the momentum of Alan Smith's system (recalled by Cohen (2009)), we are following a spiral which considers that the more society becomes rich overall, the more the division of labor increases (as currently the fragmentation of industrial production) to increase productivity and growth. There is thus a change in engineering work: in the past, following the example of Gustave Eiffel, we could, relying on the competence of a person (and at least part of his team), participate in technical progress. Today, this progress requires the convocation of many disciplines (e.g. for additive manufacturing which will be discussed in the second part (Chapters 2 and 3 of Volume 2) of this work in two volumes: materials, process, printability, functional specifications, HSE, software, waste management, etc.), which can be linked together by digital systems (Sadin 2021).

Boly et al. (2016) point out that innovation, the result of a validated invention, retains an indeterminate character. They write, "innovation is also defined by its organic and constructivist aspect: it is initiatives to adjust to constraints or to adopt new practices that create innovation", and propose a "complete" model of the innovation process, as shown in Figure 2.23.

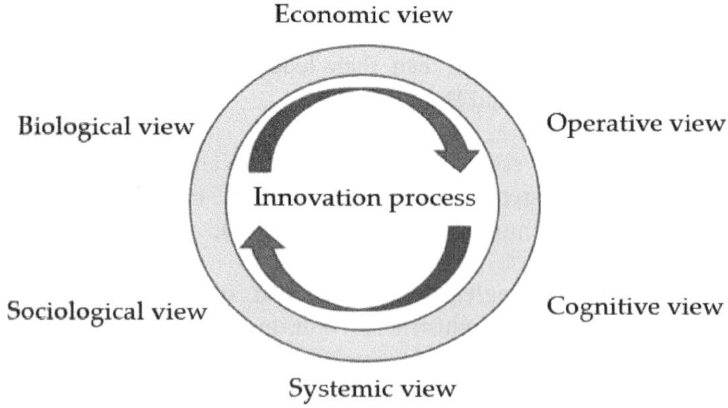

Figure 2.23. *General views on innovation*

Groves (2017) thus argues that anticipation and bricolage are practiced in different ways and depend on non-standardized elements and processes that intertwine with these practices. It is therefore a question of articulation and integration between different, partially disjointed components that it is possible to judge whether technologies are becoming "mature". The foundations of successful innovation must take into account a number of factors, as shown in Figure 2.24.

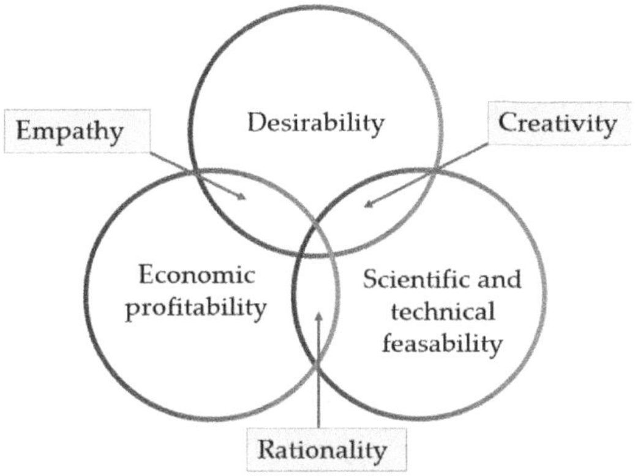

Figure 2.24. *The foundations of successful innovation*

Communities with a shared disciplinary passion are innovative, but sometimes tend to close in on themselves. So, from a scientific point of view, we need to be able to develop an interdisciplinary, heuristic, temporary and open research space, allowing a more collective approach to the creation and exploitation of new concepts, because the demand is there. Thus, in association with scientific disciplines, on subjects of public utility, we need to bring out discontinuities inducing scientific recompositions and adaptations.

To this end, all scientific and technical avenues must be exploited to satisfy an intentional research objective on material or nonmaterial "objects/devices":

> Either the researcher treats the object in a partial way, ignoring or glossing over all the links that inevitably exist between this partial view and the global biases he cannot fail to have on the object. Or else, he places his global biases at the heart of scientific questioning: not as interesting elements to look at, but as the motors of a dynamic that then takes the inevitable desire for unity as a stepping stone towards what becomes a demand for unity (Alvarez-Pereyre 2003).

This desire to get closer to reality, with all its difficulties (Nicolescu 1996), must define a common cultural background, that of modeling and integration (systems approach), combined with that of experimentation (systemic approach), combined with experimentation, as well as an indispensable openness to other scientific and technical fields, by "borrowing" their concepts, methodologies (principles of constant learning and cross-fertilization), their instruments, by "bringing in" subjects for study, and so on. All disciplines are involved from the "hard" sciences to the humanities and social sciences.

NOTE.– Definitions of the invention:

Noailles (2011) has selected a number of them and presented as follows:

– In 1995, for the European Commission "[innovation] consists of successfully producing and exploiting novelty in the economic and social spheres. It offers new solutions to problems, thus helping to meet the needs of individuals and society".

– In 1997, the Oslo Manual states: "A technological product innovation is the implementation/commercialisation of a product with improved performance characteristics such as to deliver objectively new or improved services to the consumer. A technological process innovation is the implementation/adoption of new or significantly improved production or delivery methods. It may involve changes in equipment, human resources, working methods or a combination of these".

– In 2008, the USDC defined innovation in its report as: "The design, invention, development and/or implementation of new or altered products, services, processes, systems, organizational structures or business models for the purpose of creating new value for customers and financial returns for the firm".

– In 2009, the European Commission, linking invention and innovation is the ability to take new ideas and convert them into commercial results more effectively and rapidly than the competition, thanks to new processes, products or services.

– In 2010, the OECD drew on the Oslo Manual to propose: "Innovation is defined as the implementation of a new or significantly improved product (good or service) or process, a new marketing method or a new organizational approach in business practices, workplace organization or external relations" (OECD and Eurostat 2005).

– Finally, Noailles proposes the following definition, which will be retained: "An innovation is the implementation on a macro-economic scale of a new technique, a new tool or a new organization in the broadest sense of these terms, leading to a lasting improvement in the overall efficiency of society as a whole".

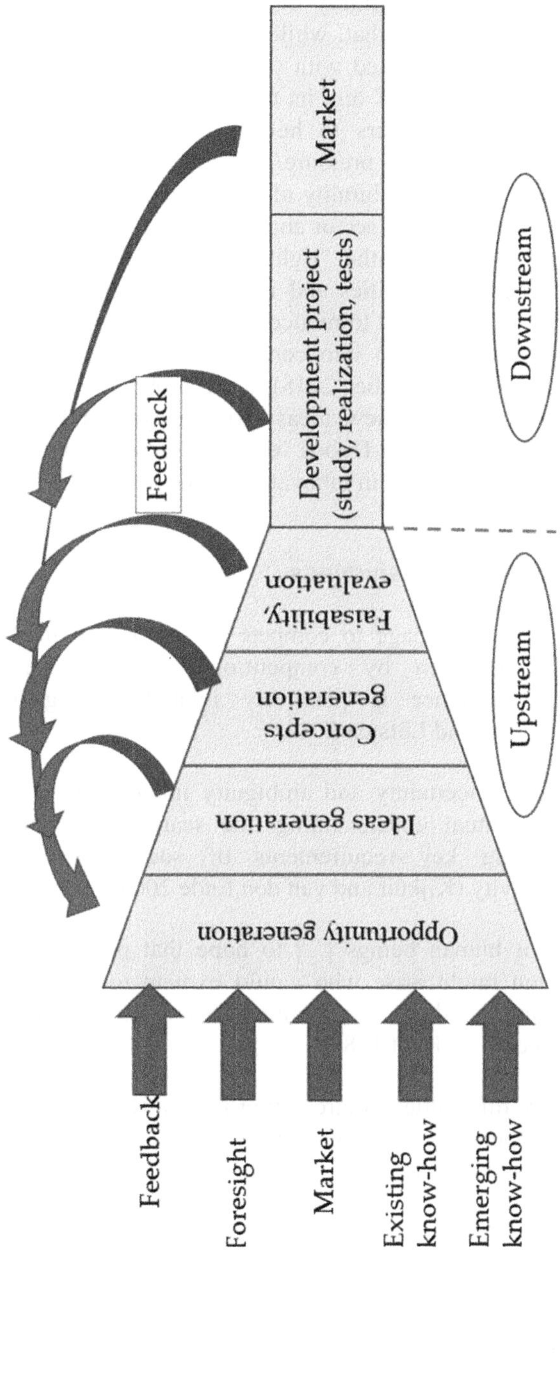

Figure 2.25. *Schematic representation of the innovation process*

These definitions simply reflect the inevitable link between invention and innovation. But, in fact, a plurality of players are involved in this process. Charue-Duboc et al. (2020) show that, while one pressure is exerted to encourage invention, there is another associated with the many experts involved in the long journey between the validated POC and its translation into the field. By involving more and more specialized players to help with the transition from POC to application, by increasing the time pressure on them, by requiring feedback to the invention, and by implementing a plurality of phases depending on the nature of the design activities, the steering mode sector and managerial decisions, it is possible to imagine the difficulty of making the "right" decision. For Charue-Duboc et al. (2020), the link between communities and innovation is complex, multiform and multidimensional and, in an attempt to reduce the complexity presented above, three types of contribution can be taken into consideration: the capitalization of good practices (Goglio-Primard and Soulier 2018), agile problem solving (Guillaume et al. 2017) and the generation of new ideas (Arzumanyan and Mayrhofer 2016). Figure 2.25, taken from Charue-Duboc et al. (2020), provides a schematic representation of the links involved in innovation processes.

2.6.2. *Innovation and responsiveness*

> How do companies manage to compete in a turbulent market, and avoid being overtaken by competition? [...] We show that organizational resilience is positively related to organizational creativity (Richtner and Löfsten 2014)[12].
>
> The complexity, uncertainty and ambiguity inherent in the creative process make mutual understanding, the search for meaning and consensus-building key requirements of social interaction for successful creativity (Kijkuit and van den Ende 2007)[13].
>
> It is absurd for human beings [...] to hope that perhaps some day another Newton might arise who would explain to us, in terms of natural laws unordered by any intention, how even a mere blade of grass is produced [...] (Kant 1987).

However, there are difficulties in circulating innovative ideas and projects from exploration units to operating units (O'Reilly and Tushman 2004), as there are compartmentalization and closure effects, which tend to increase with company size.

12 Author's translation from the French version of the text.

13 Author's translation from the French version of the text.

For Charue-Duboc et al. (2020), these filters can raise questions about the scope for innovation in these structures. Indeed, any invention has to convince each layer of the decision-making process, which operates according to its own criteria "that go far beyond the sole scope of technical feasibility or pertinence. Coupling issues, particularly between technological and usage dimensions, are key factors here, when it comes to proposing new offers, based on distinctive technologies and featuring value-generating functionalities" (Charue-Duboc and Gastaldi 2017).

Aimé and Arnould (2021) comment on the notion of ontology, which could be applied to the upstream–downstream domain of innovation. This would involve attempting a conceptualization, that is, a representation of a simplified vision of the world, depicted as in Figure 1.9, with the aim of better understanding the links involved and their mutual importance. The meaning of concepts modeled in this way should then be associated with interrelations and interdependencies. To the best of the author's knowledge, this modelling to the author's knowledge has not gone beyond this summary description, but it would be of interest in moving forward on this complex dynamic.

Independently of this aspect, with international competition, the world of organizations and that of decision-makers must satisfy a reactivity, a real risk-taker, but also an almost absolute. In this context, it is essential to avoid the trap of "technological determinism" (Chardel and Dartiguepeyrou 2018). Of course, it is possible to save time thanks to the digital society, but we must avoid a reductionist approach that systematically destroys anything that can slow things down, because this can be detrimental to (good) decision-making. But when it comes to decision-making, a relationship to acceleration, and more generally to time (which is not uniform and differs according to individual values and corporate cultures), needs to be mastered. In this process, it is not a question of playing with passive filters, but of supporting motivation and mobilization of the people responsibly affected by the decision.

Let us remember that to be passive or demobilized is to accept anesthesia in the face of economic consequences. The size of the network of intermediate and final decision-makers can obviously be an important inertial factor; this reality therefore militates for the creation of flexible, adaptive structures that can escape delays and misunderstandings; these are startups. And then, as often happens, in case of success, nothing prevents the large company from buying the startup, either to stay in its traditional field and avoid competition, or to integrate it into its own dynamic. Figure 2.26 concerning startups recalls the critical elements to consider in their implementation (Tripathi et al. 2018).

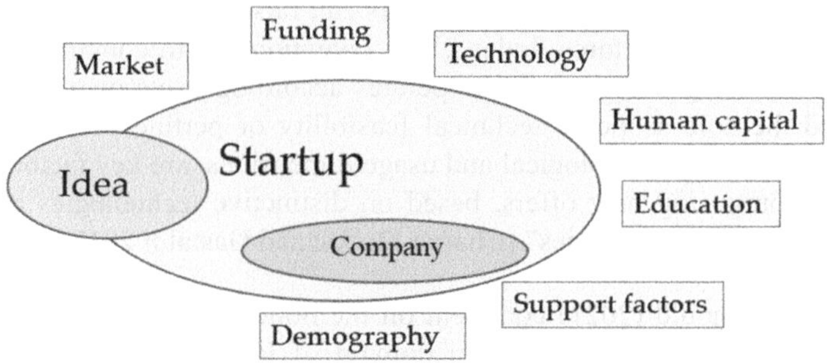

Figure 2.26. *Critical elements associated with a startup*

NOTE.– Startups and large companies:

An analysis by Stacey et al. (2021) revealed that large technology companies are buying up smaller competitors at a high rate. According to this source, technology companies have spent at least US $264 billion on potential competitors. This surfeit of acquisitions is taking place against a backdrop of intense scrutiny by the US government for its desire to stifle competition and harm consumers. Moreover, since transactions valued at less than $92 million do not have to be reported to American regulatory authorities, an analysis by the *Financial Times* shows that since the beginning of 2021, technology companies have signed 9,222 deals to buy start-ups worth less than a billion dollars (around 40% more than in 2000). The Biden administration is currently seeking to combat the concentration of corporate power in several sectors.

NOTE.– Support for "unicorns":

France is aware that it is lagging behind in the creation of innovative start-ups; it has implemented a proactive strategy to channel the resources, particularly financial, needed to ensure their success. There are currently 18 French unicorns on French soil. The Institut Montaigne (2021) has identified the success (or failure) factors that contribute to the emergence of resilient, high-performance ecosystems. These ecosystems develop comprehensive asset-based strategies that leverage economic, physical and social networking forces. The Institute proposes recommendations to make France the most innovative country in Europe, and to enable it to create the most high-value emerging companies:

– pursuing efforts to finance innovation in particular by encouraging the French to direct their savings much more toward supporting start-ups;

– investing in human capital by piloting an annual national exercise to determine skills needs over the next 10 years, adapting higher education training accordingly, and broadening the pool of business creators to include as many talents as possible;

– strengthening the links between research and entrepreneurship by implementing an ambitious promoting and facilitating the creation of innovative companies by researchers in public laboratories and by developing a competitive project selection culture modeled on the (DARPA: Defense Advanced Research Projects Agency).

NOTE.– Startups and universities:

Students can play an important role in the entrepreneurial dynamics of universities (Shah and Panhke 2014; Wright et al. 2017). They are often the holders of the know-how needed to exploit patented inventions in academic spin-offs (Hayter et al. 2016; Schaeffer et al. 2018; Bolzani et al. 2021). At the same time, they create more companies than academic researchers (Siegel and Wright 2015). As a result, a number of universities have become involved in building entrepreneurial ecosystems adapted to local contexts and fostering the development of an entrepreneurial culture through teaching and support for student entrepreneurs (Wright et al. 2017; Matt and Schaeffer 2018). Each university may have its own entrepreneurial model (Schaeffer 2019).

Across the Atlantic, one solution to overcoming the "valley of death" was, according to Kunegel (2019), the creation a "Proof of Concept Center" to support spin-offs. In particular, it is possible to facilitate access to funding information asymmetries (Eschenfelder and Shankar 2017). In this context of real and verifiable aid, researchers feel more concerned about promoting their work (as in France, no hierarchical pressure is exerted).

NOTE.– Globalization:

International trade plays a role as a channel for technology transfer. The granting of licenses has a positive effect on their participation in bilateral trade and in their inclusion in global value chains, boosting corporate performance. The acquisition of foreign technologies would lead to significant improvements in productivity (Santos Gomes and Kovaleski 2019; Rigo 2021).

But also, we can have an anti-fragile system, which must be non-predictive by definition. With the avoidance of clones and one-track thinking in the decision-making system, which must be "agnostic" about the future. Based on the purpose of innovation, the decision-maker simply has to consider first the means, then the ends. But it is not their job to resolve possible disagreements around a common belief, and so they have to leave the various options open until the final decision. If the dominant option fails, other options become available. It is therefore a method of collective complex problem-solving through free experimentation with possible options (see Taleb 2013). For his part, Noailles (2010) proposes the structuring between upstream and downstream sectors shown in Table 2.4. However, in this method, a central distinction must be made between science and technology (see Chapter 1 of this volume) as repositories of knowledge enabling invention and engineering methods as the dynamic processes by which artifacts are created.

A creative contributes his vision via a POC, which is taken up in a wider context to produce a DI that must satisfy social criteria. Snider et al. (2016) then advocate a process of expert reflection to reframe the context and examine whether an even more original solution can emerge. Here, forcing an unusual starting point encourages the development of an unusual solution.

Boden (2004) and Wiggins (2006) argue that creative behavior can occur through the direct exploration of conceptual space, or through the transformation of conceptual space and the rules that describe it. Creative designers work from first principles rather than from a frame of reference based on experience; so, starting with unusual examples and perspectives can produce more creative results (a process called de-fixation). Another possibility is to assume that the project is an "ill-posed" question, even when a well-defined structure is available. This forces one to form and follow an initially unknown or unusual process, which leads, according to these authors, to the development of a creative solution.

Function	Thinking	Inventing	Innovating	Engaging	Producing
Activity	Research	Development	Marketing	Production	Mass production
Productions	Ideas, articles, books, software	Model, prototypes, tests	Standardized products	Series products	Social standard
Actors	Academic researcher	Inventor, enginer	Innovator	Entrepreneur	Manager
Environment	Academic research	R&D center, startup	Compagny	Capitalized company	Multinational
Funding	Public	Start-up	Risk-capital	Development-capital	Loan

Table 2.4. *Upstream–downstream relationships in innovation (disruptive)*

But to smooth the transitions between upstream and downstream, and thus combat a number of conservatisms that are sometimes elevated to the level of ethics, we can use fiction to appeal not only to the imagination of creative minds, but also to society as a whole (and thus to decision-makers). This is undoubtedly a useful approach to the emergence of DI. In this context, Blanco (2008) argues that innovators are aware of the weakness of their predictions, but that this serves to rationalize them necessary to convince potential partners (see also Kirzner 1979; Weick 1995). This proposition assumes that the creative person has the time and interpersonal skills to create a space of precomplicity that facilitates innovation.

As shown in Figure 2.27 (OECD 2013), the technological component can be broken down into proprietary technology (the black box), technological platforms and "infra-technologies". Proprietary technology with its R&D is applied to industrial development, while the technology platform refers to the POC leading to the ID, while infra-technologies are a set of tools and methods used by the high-tech industry (including standards).

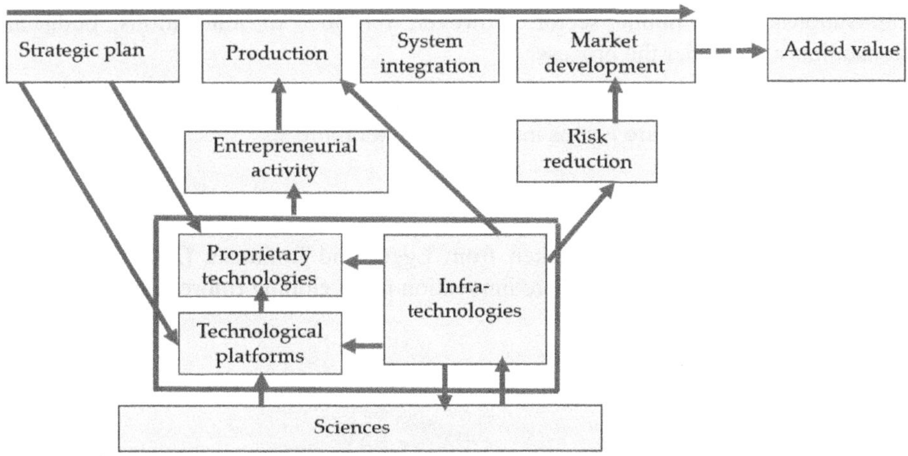

Figure 2.27. *Industrialization*

NOTE.– Myths about innovation:

According to Apolitical (2019), there are several myths about innovation:

– Myth 1: The public sector is not innovative. The public sector is vast, cumbersome and bureaucratic. Four out of five (80%) public establishments introduced at least one innovation between 2015 and 2016.

– Myth 2: The private sector is more innovative than the public sector; 44% of Danish companies were innovative in 2016.

– Myth 3: Civil servants oppose change. The public sector is overrun by conceited specialists, creatures of habit with an employee mentality, powerful unions and collective agreements that regulate everything. Civil servants play a key role in 90% of innovations in the public sector.

– Myth 4: The public sector spends its time reinventing the wheel. Only 18% of innovations are new.

– Myth 5: The public sector is afflicted by the "not invented here" syndrome. Only 15% of establishments directly copy other people's solutions.

– Myth 6: Digital platforms are the most effective way of disseminating innovation. Only 11% of innovators said they had heard about the solution on websites and in e-newsletters, 8% on social media and 8% in the media.

– Myth 7: Financial pressure is always good for innovation. Financial pressure can actually promote the innovation process. It is a driving force for 23% of innovations in the public sector. However, for 36% of innovations, budgetary constraints also hinder the process.

– Myth 8: Denmark is a pioneer of innovation in the public sector. All of the other Nordic countries are just as innovative, if not more so.

It would be interesting to carry out the same survey in other Western countries.

Furthermore, Figure 2.28, taken from Egger and Loumeau (2019), shows the European geographical areas where innovation is developing (more than others).

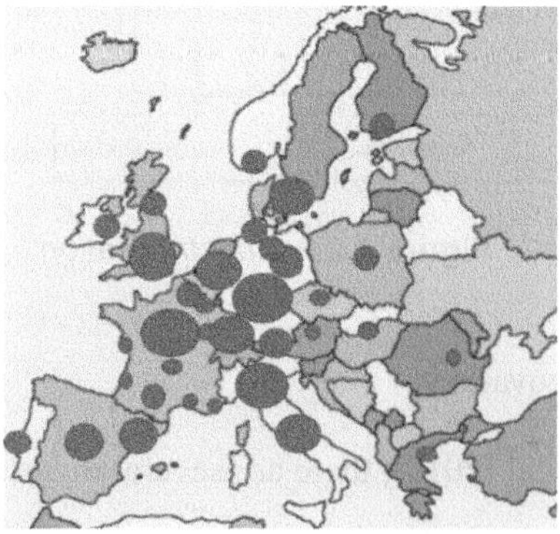

Figure 2.28. *Who innovates where?*

Van de Vliert and Murray (2018) show in Figure 2.29 the influence of latitude on creativity (and therefore of temperature on this same criterion) through their analysis in 155 countries. What effect will global warming have on innovation?

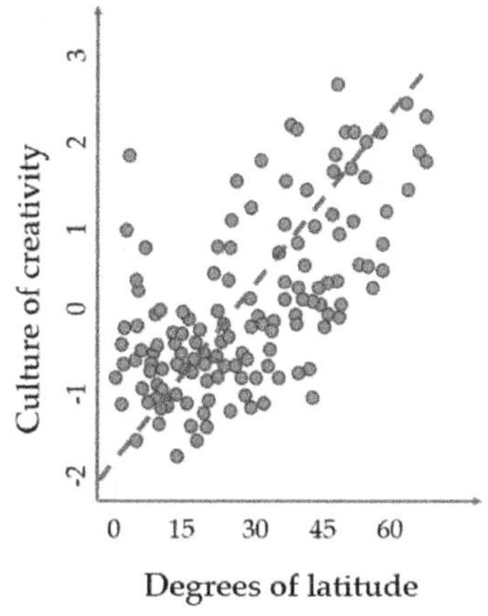

Figure 2.29. *Creativity correlation and latitude*

2.6.3. *Innovation programming*

Innovation diffusion is not a rational implementation process, but more accurately portrayed as a highly social process, involving sets of intermediate organizations that contribute a product's reputation (Recker 2016).

Opinion leaders are identified as an actor to whom others turn for advice on complex situations, and are therefore seen as expert and credible. They are well-connected actors with extensive social and peer networks (Caiazza and Volpe 2015)[14].

Creativity is the cornerstone of scientific thinking, but scientists in academia are judged by metrics of quantification that often leave little room for creative thinking. In many scientific fields, reductionist approaches are rewarded and new ideas viewed skeptically (Angeler et al. 2016).

Valley of death no. 10: getting there in time

The OECD (2021a) has defined coordinated sets of policy and regulatory measures to mobilize science, technology and innovation within a programmatic framework. These measures cover different stages of the innovation cycle, from research to demonstration to market entry, combining initiatives to "push" the

14 Author's translation from the French version of the text.

market by acting on the supply side, and "pull" it by acting on the demand side and cut across various sectors and disciplines. The characteristics expected of MOIPs (Mission-Oriented Innovation Policies) are defined according to three dimensions:

– strategic orientation: legitimacy, direction, intentionality and flexibility;

– coordination policies: horizontality, verticality, intensity and novelty;

– policy implementation: coherence of policy mix, financing capacity, evaluation and reflexivity.

This organization has identified some 40 policy initiatives that correspond to these major programs based on the following elements:

– mission-specific global strategic frameworks;

– challenge-based programs;

– ecosystem-based mission programs;

– mission-driven programs.

Thus, the main results proposed in OECD (2021a) are as follows:

– most programs follow an open, non-prescriptive approach in which "problems are selected, not solutions". However, few of them consider social innovation;

– for some programs, objectives with the expected mission characteristics have been set: clear, bold and inspiring, with high societal relevance, targeted, measurable and limited in time;

– missions are generally not defined at the outset: rather, they are the result of a very gradual and inclusive process;

– almost all programs combine social and economic objectives;

– all programs are directed and managed by a complex, multi-level governance structure;

– there are very few evaluations of these programs;

– although difficult to estimate, most programs involve additional costs;

– no program has been launched from scratch;

– program design is strongly influenced by the specifics of the national institutional framework;

– there's virtually no room for break-ups.

As has been widely presented, there is room for increments, but little for more "revolutionary" actions, when we know that they constitute part of the solutions expected by society. For example, the EU's most ambitious innovation initiative launched by the European Innovation Council (EIC), part of Horizon Europe, a heavyweight operation with a fund of €10 billion to provide grants not dilutives and equity investments in young innovative shoots (startups) within the EU (CE 2021). However, MKGI (McKinsey Global Institute 2019) estimates that over the last century, Europe has largely lost its advantage in the field of innovation (Europe spends the most money on research and development (R&D) of public origin in the world, its private investments represent only 19% of the world total, behind China (24%) and the United States (28%)).

Moreover, in France (but not only), the trend is to support large research teams. Small research teams, or even lone researchers or inventors, are tending to disappear. This situation may lead us to believe that the choice made by research ministries is in the interests of future scientific progress. However, according to Wu et al. (2019), this is a false impression: large teams contribute to the advancement and development of science, while small teams often produce results that disrupt science.

Smaller teams are more likely to find results that change the way people think, or a complex problem that can then be tackled by a larger team. They would be more inclined to dig into obscure ideas from the past to kick-start their research. Furthermore, for the Institut Montaigne (2021), the worlds of teaching, research and business are still too compartmentalized in France. This lack of diversity is also reflected in the low number of international profiles: 16% of startup founders in France have studied abroad and 3% are of foreign nationality. In the corporate world, the proportion of foreign researchers is 6%.

2.6.4. *Innovation positioning*

Disruptive technologies and associated business create different types of markets, present radically different challenges for established companies, and also have radically different implications for managers. These new business models make it possible to exploit under-utilized resources, bring new products and services to market, and facilitate the best trade-off between supply and demand, accelerate innovation, stimulate demand and energize the market, and foster self-organization, etc. To provide a picture of this innovation, Zubizarreta et al. (2021) propose a mapping of a number of the characteristics associated with these emerging phenomena in Figure 2.30.

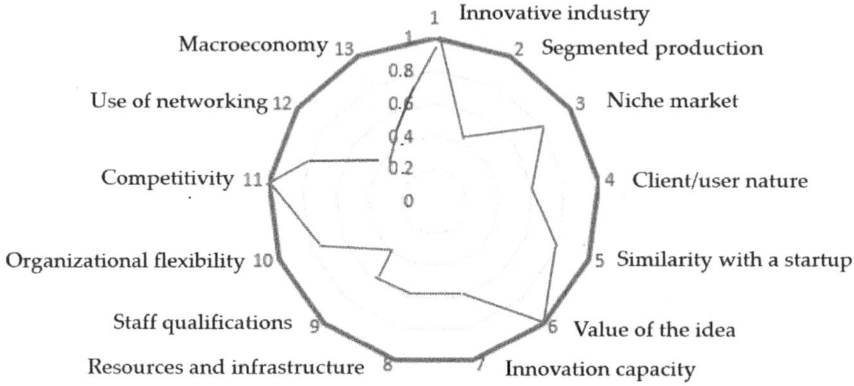

Figure 2.30. *Mapping a disruptive innovation (in red, an example)*

Figure 2.30, which only leads to qualitative indications, makes it possible to judge the relevance of the designation of innovation (new product or service) of a revolutionary or discontinuous nature, taking into account its potential to generate wealth while disrupting the previously established industry or market. Remember that DIs help organizations gain a foothold in new or low-end markets that had been ignored by incumbents.

2.7. Standards, standardization, various filters

> We realized that our observation that France had become a bureaucratic prison [...]. The bureaucratic suffering experienced in France is colossal. Even civil servants complain about it, especially those in the field (Koenig 2022).

> For better or worse, "research" has overtaken the need to summarize knowledge in a system. Rationalizations are far from having disappeared from the contemporary world; rather, they have been disseminated in various fields of experience technology, economics, language, politics (Fœssel 2015).

> Although it may seem fixed, at least momentarily, it [creativity] is in fact constantly being reshaped, redefined and re-evaluated as a function of individual actions, which may contradict, complement or surpass it (Cina 2021).

According to Cina (2021), there is a tension between economic and creative issues in organizations. It is even widely accepted today (Lampel et al. 2000; De Fillippi et al. 2007; Leclair 2018). As in the case of the Covid-19 pandemic, economic issues are taking precedence over innovation.

Furthermore, according to Holford (2019), the necessary standards for companies to work and exchange are increasingly ubiquitous and are likely to attack the creative spirit of operators and other partners in companies. According to France Stratégie (2022a), France will need to attract new production sites using new technologies to rebuild the country's industry. It points to the structural factors of attractiveness: taxation and production factor costs. It plans to identify "the obstacles related to procedures, regulations and support for the various stakeholders". Thus, the French government's concerns include the issue of new technologies (but not the means to invent and develop them) and administrative delays. For those who invent and seek to innovate, this context is a "valley of death".

2.7.1. *Norms and standards*

Western society made substantial progress when it was able to standardize the production of its weapons. From the manufacturing and repair points of view, there is a development of knowledge and understanding for rationalization and standardization in all its forms. This standardization involves a process of unification through "standards" and norms of products (Busch 2011), procedures and forms of administration (Gardey 2008). For Ruano-Borbalan (2017a), it is a unification of intellectual or scientific content, a formatted way of thinking. The massification of everything can then be associated with impoverished thinking. According to the same author, the consolidation of standardization is part of a dual movement: the historical development of the nation-state and the hold of administrative frameworks and standards (Bush 2011; Beze 2014), useful scientific development, but constrained for many reasons by quality approaches (scientific procedures, repeatability of results, avoidance of plagiarism, management of instruments, performance-based management instruments or the organic law on finance laws (known as LOLF) of 2001, etc.):

> The standardization of knowledge in the very contemporary period is increasingly focused on the normative codification of individual behaviours and "skills". This standardization of social and professional behaviors is the precondition for counterpoint discourses promoting creativity and individual identity (Ruano-Borbalan 2017a).

For Meyer and his co-authors (Meyer and Rowan 1977; Meyer et al. 1987, 2006; Meyer 1994), a dynamic of rationalization through specialization and the quest for efficiency and effectiveness is a global dynamic that characterizes what they call *world polity*. According to Towley (2008), bureaucratization, centered on the optimal matching of administrative means to stated ends, leads to an "illusion of scientific objectivity", with the concept of "performance management". This rationalization, which assumes that we are capable of knowing precisely the input and output data of a system that must be deterministic, manifests itself through a wide range of instruments: systems for measuring objective-based performance measurement systems; reporting processes, benchmarking, rankings, contractualization systems, cost accounting, performance-based budgeting, internal control, etc. (Bezes 2014).

Even the author was appointed to try to define in 2014 also, for CNRS research units, measures of the efficiency of these research organizations. Including some complex aspects, this work (as the author wished), which could be of interest in terms of reducing costs and headcount, disseminating performance tools (what performance?), even reducing autonomy, etc., was not retained by management, no doubt because a software package integrating the minimum of data could not, on the proposed bases, produce an efficiency figure. Among the concrete consequences of the rational myths, let us mention the implementation of the general review of public policies (RGPP), stemming from the New Public Management, with its emphasis on the quantitative.

NOTE.– Standards and standardization:

In industry, the economy or services, a standard is a set of conformity or operating rules legislated by a mandated standardization body, such as ISO (International Organization for Standardisation), while a standard is a set of recommendations or preferences advocated by a group of characteristic, well-informed users (a form of destroying diversity in favor of facilitating to facilitate exchange between companies and contribute to the stability of systems). Standards enable rapid adoption of technological innovations. Standards enable the transition from a "cottage industry" to an industrial scale. But this takes time. In just a few decades, international standardization (ISO) has become a decisive regulatory mechanism in our societies, with several tens of thousands of standards (Ruwet 2017). ISO 9001 defines the criteria applicable to a quality management system, the only standard that can be used for certification, which means rigor in stabilized production.

"The dissemination of myths is therefore neither magical nor automatic: actors do not seize upon them mechanically. Their legitimacy is not self-evident, and they have to be adapted, not threatening well-established power situations, and sometimes relying on coercion" (Bezes and Musselin 2015). But, as we have seen, these organizational transitions are also spaces for the expression of power relations and power struggles based on the maximum number of quantitative judgment criteria in the name of a principle of rationalization. As in Silicon Valley, where creative people are found in large numbers in knowledge-based activities, there is a sufficient critical mass of ideas to sustain the technological performance of firms based in this area. These are fast-moving, "chaotic" markets, where there are enough start-ups to define other standards more suited to radical innovation. They move away from the aforementioned modes of operation of the state and "oligopolistic" companies.

2.8. The question of industrial disruption

> Even though they are competent, honest and dedicated to the common good, some technocrats [...] give me the impression that they are incomplete, boring and stupid human beings [...]. They are so afraid of the fear of others and of their own fear that they have never sought to understand it, nor to see in it an unsurpassable fact of the human condition (Dupuy 2018).

> Anticipation isn't just forecasting; it's much more than predicting what the next event will be. It's more than coping with the next event, it's creating the next event (Follett 1949)[15].

> The acceleration of the pace of life made the people of the 19th century realize that they were immersed in an absolute flux where every stable entity was revealed to be a fiction, fragile and provisional. This revelation brutally destroyed all the reassuring support of beliefs and reference points that ancient societies had built up for themselves (Stiegler 2021).

Valley of death no. 11: reaching the market

At this stage, this means that the company backing the operation has to make a trade-off between investment and expected return. Several issues need to be taken into consideration: the difficulty of industrializing the innovation at an acceptable cost, inadequate training time in relation to production, lack of success due to a lack of "desirability" because it arrived too early or too late, a market that exceeds what the company can do and so on. Between the pluses and minuses of profits, speculation and risk-taking, the return on investment is unpredictable.

The world we have been living in for a long time now is based on two main pillars: a system of production and a system of organization. The former covers the

15 Author's translation from the French version of the text.

following elements: creation and sharing information, food, transport, energy and materials. It must satisfy the material needs of a society. These elements are naturally interdependent. The second evolves at the same time as the first and challenges the way it governs social behavior (thought cultures, beliefs, economic models, governance structures, etc.). It is the key element for decision-making:

> The decisions we make at the organizing system level will therefore be crucial in determining the ultimate design and possibilities of the technology-driven societal disruptions currently underway (Arbib et al. 2021).

"Politics is not a matter of *épistémé* [science], but of *doxa* [opinion] [...]. Not all doxas are equivalent, and there is a kind of knowledge in politics that is not 'science', but a matter of judgment, prudence and plausibility" (Castoriadis 1999). Herbert Simon, winner of the Nobel Prize in Economics in 1978 (quoted by Jarroson (2017)), observed that, since everything cannot be known, decision-making is based on incomplete information about possible options and their consequences. Often, the decision-maker does not seek the optimal decision but settles for the first acceptable decision for them and their immediate environment.

"We are in the midst of a long and painful process leading to the emergence, in one form or another, of a global society whose probable structure it is not yet possible to imagine" (King 1999). Recent trends, authorized by the decline of a certain balance of terror during the Cold War, and likely to be revived by current "European wars", are reflected in the multi-faceted phenomena of a Europe-wide association, with a globalization of trade on the one hand, but also environmental concerns, the reemergence of disjunctive forces, disenchantment, irrationality, xenophobia and other forms of fundamentalism, inward-looking attitudes, competition, etc., on the other hand. In short, the future mapped out by the liberal system, in which peace is a condition for "living together", is no longer as secure as it was yesterday, which means that we need to look at possible strategies that condition and are also conditioned by the digital world which can no longer be ignored. In fact, social structures are innervated by new, immediate digital information channels, suffer from exclusion, no longer have a secure collective vision, and see their values modified along with their culture, sometimes reaching the point where they can no longer be ignored with their culture, sometimes reaching a sterile uniformity. In the space of a few decades, many self-evident principles have become obsolete.

Within organizations, there is a whole range of more or less eclectic "sticking plaster" approaches (increasing costs and diminishing returns) for dealing with problems when a technology from subsidies subcontracting solutions to the production of technical products at lower cost abroad. In effect, we have only treating the symptoms, not the root of the problems linked to the technical performance of production. A simple and effective approach would be to focus on key technologies that could transform the foundations of the industry. This angelic comment does, however, call for a few observations:

– other industrialized countries can make the same analysis, and the competition we tried to avoid is likely to recur;

– it is unrealistic to assume that disruptive ideas of sufficient quantity (and importance) can be generated in a short space of time, especially if the framework for supporting invention is biased;

– ensure open, competitive and transparent markets (limit regulatory barriers, avoid blocking monopolies, etc.);

– we need to anticipate the need for appropriate training for those who could support the future industrial effort based on disruptive ideas.

2.8.1. *Disruption conditions*

> Technique exists, not as a docile servant applying intangible orders, but as [...] drawing attention to concrete problems and directing research towards obstacles without presuming anything in advance about the theoretical solutions given to them (Canguilhem 2013).

> Evolution, then, is in all cases a change from a more diffused or incoherent form, to a more consolidated or coherent form (Spencer 2017).

> I explained in detail in my Dissertation on the Principle of Reason how it is the condition of every possible object; that is, any object is necessarily linked to others, being determined by them and determining them in turn. This is so true that the whole reality of objects, as objects that are nothing but representations, is reduced to this relationship of necessary and reciprocal determination; this reality is therefore entirely relative (Schopenhauer 1997).

What we call "science", in the sense of pure, disinterested knowledge, was officially "discharged" in 1637, when Descartes' *Discourse on Method* was published by the "master and possessor of Nature". This possession (under threat for a variety of reasons, including global warming and the depletion of reserves) must not separate the viewpoints of scientific "professionals" from those of society as a whole, for whom they must exercise their talents. It is not easy to move away from secure knowledge within solidified paradigms, and yet there are creative people who imagine a future other than the one that may unfold by successive increments. These are the creative people who, after a long road, lead to revolutions, large or small, that change the world. They are briefly mentioned here, in this section devoted more specifically to the industrial aspect.

For Schumpeter (1984), innovation and technical processes are the main driving forces behind economic progress. New technologies are emerging faster and faster, and with global relocation, many developed countries are losing parts of their economies. At best, they replace new economic opportunities (idea of creative destruction). In the Schumpeterian model, the entrepreneur is at the heart of the capitalist system, with their dynamism (unlike Keynes) breaking the stationary or conservative state of the economy. Growth is integrated into a permanent movement of creation, destruction and restructuring, in which entrepreneurs are more motivated by creativity than by greed alone.

2.8.2. *Local changes, systemic changes*

> The reality is quite different. Every invention allows us to solve old problems, but new ones immediately appear, in a different field. We are caught up in a movement that, far from slowing down, is accelerating. Let's not expect us to be able to make a few efforts and then rest. Our very progress requires us to seek further (Berger 1964).

> The old technology imposes its own standards of economic evaluation on the new technology, elaborated with reference to its natural qualities, thereby introducing a kind of bias at the time of the economic calculation (Foray 1992).

> Every profession makes progress, but progresses in its own groove [...]. The groove prevents straying across country, and the abstraction abstracts from something to which no further attention is paid [...]. But the point is the restraint of serious thought within a groove. The remainder of life is treated superficially, with the imperfect categories of thought derived from one profession (Whitehead 1975).

Apart from these divergent random events, if it is possible to understand the main drivers and levers of major systemic change, these few comments show that there are many obstacles to overcome in terms of creativity (Roquette 2007), the orientation of inventions, training, the materials to be supplied, etc. A pedagogy of adventure defined in terms of the quality of life associated with technology, a form of storytelling, encourages citizens to experience the unknown, offering them the opportunity to draw analogies between distant fields of knowledge (science and art, technology, energy and culture, etc.). With the development of information reduced into an increasingly binary presentation, the citizen is increasingly doomed to uninformed ideological recourse. And technology alone is no longer enough.

Nevertheless, Arbib et al. (2021) consider that paths to success associated with disruption are possible when "different actors seek to accelerate disruptions in their own context to avoid risks or secure benefits, the ensuing race to the top will generate further powerful accelerating feedbacks that will exacerbate the speed and scale of disruption on a global scale". Indeed, conventional scenarios are unrealistic because they are based on a fundamental misunderstanding of the transitions enabled by disruptions: there are numerous examples, such as those linked to machinismo, concentrated energies, automobile travel, electricity, electronics and the Internet, or quite simply 3D printing, examples that demonstrate these evolutions through qualitative and quantitative leaps.

Technical change, as it is perceived by many, has a number of specific characteristics: it is irreversible, irresistible, discontinuous and uncontrolled. Unnoticed by many, this rapid transformation of the production system disrupts the culture of society as a whole, creating new hopes and new possibilities, but at the same time new alienations, even new real or merely perceived risks.

For these three authors, technological disruptions occur between periods of technological stability and can trigger brutal economic and social transformations:

> Disruption occurs when a new technology of equal or greater capability becomes available at a significantly lower cost than existing alternatives, after which the incumbent technologies are replaced very swiftly, with the market share of the new technology rapidly growing from 10% to 90% or more, often within as little as 15 years (Arbib et al. 2021).

Emerging technologies then have the opportunity to expand their markets, create new business models and generate entirely new markets. They also lead to new learning and new habits.

The convergence of new technologies, such as the concept of Industry 4.0, can trigger synergies and feedback loops within sectors, as explained in André (2019). He writes:

> History shows that these loops interact and amplify each other, accelerating the adoption of new technologies in a virtuous cycle while accelerating the abandonment of old technologies in a vicious cycle.

The automobile has not simply replaced horse-drawn travel, photography, drawing and so on. Disruptive technologies do not simply replace existing technological modes, but open up radically new uses, sometimes unanticipated by their inventors (e.g. the famous "pink" Minitel). They can appear when several technologies converge to enable the development of new products or services that are even more effective than existing ones.

However, Figure 2.31 is a reminder of the obvious: the allocation of resources to research and R&D cannot go beyond a certain limit, for a variety of reasons: it's arithmetically impossible to exceed GDP need to train researchers, research and R&D are only part of the overall innovation process from invention to market launch, etc. And then, for political reasons, there may be other priorities. But in the 1960s, major impetus for the development of scientific research in France led to the belief that this age of support could continue for a long time to come and violent reactions associated with financial limitations or attempts to optimize the credits allocated to public research (Demoly and André 2022).

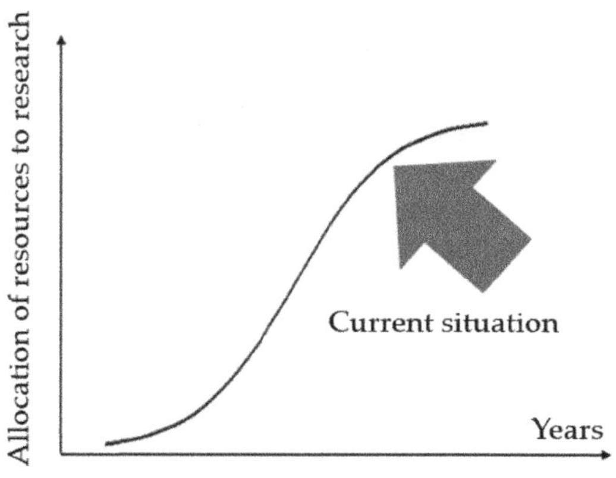

Figure 2.31. *Limiting the allocation of resources to research following the emergence of a disruption*

By promoting one or more forms of desirability, disruptive technologies can be attractive, sometimes at a higher cost (at least initially) than older technologies, because they are not simply substitutes for older products. They lead to a different vision of the (social) use of technologies (especially as in Industry 4.0 there is a convergence of technologies). Arbib et al. (2021) take the example of the transition from horse-drawn and internal combustion engine transport, where this development was associated with the development of materials (metals, rubber, etc.), assembly-line production (Taylor, Ford) supply means and organizations (gas pumps and transport); in short, with radical changes in society. It would also have been possible to take the example of the ubiquitous use of electricity, with its consequent disruptions (in the form of a chain reaction) in building (skyscrapers with elevators), work (night work with lighting), comfort, the relocation of energy uses and so on. What these two brief examples show is that (major) disruptive technologies are not confined to a single sector.

NOTE.– C-K method:

In this book, we have taken the idea to an industrial application. Hatchuel and colleagues (see Le Masson et al. 2018) introduced the C-K method which, starting with the company, creates a "background map" and selects the paths that teams will work on with a view to positioning themselves in the medium term. The approach proposed by Baldwin Partners (2019) comprises the following four stages:

– define: define project framework to select missing knowledge;

– knowledge: new knowledge required for innovative design – this work enables us to define projective concepts;

– concept: a creative activity which typically results in twenty or so concepts, which are then mapped out to create a base map of possibilities;

– plan: scoring of ideas and definition of the constructive line of improvement.

The aim for companies using the C-K method is to create a "guideline" that provides an action plan based on their specific needs and timeframe.

2.8.3. *Convinced?*

> [It] is necessary for us to be certain, that to imagination is added something which is reasoning (Spinoza 1954) OR [it] is necessary for us to be able to change things, that to reasoning is added something called imagination (Cargnelli 2020).
>
> The world's problems cannot be solved by skeptics or cynics whose horizons are limited to obvious realities. We need Men capable of imagining what has never existed (Kennedy 2005).
>
> The mind's faculty of appropriating what is foreign to it is revealed in its strong tendency to make the new similar to the old, to simplify the diverse, to neglect and repel what is completely contradictory [...], all things necessary according to the degree of its faculty of appropriation, its faculty of "digestion" to put it colorfully (Nietzsche 1979).

Writing that “it takes quite a different strength and mobility to hold oneself firmly in an unfinished system, with free and open perspectives, than to hold oneself in a dogmatic mode”, Nietzsche (quoted by Taguieff (2021)) is still alive and active in the present time. When we examine a field such as that of scientific research, we have to validate this position where, in New Public Management, it is quantitative evaluation that is rather advocated; the same is true of the operation of many companies which know that they must innovate to survive, without being able to undertake the necessary work while keeping, by continuity, forms of tradition that ignore changes (including environmental).

In part, this is why few countries are able to generate truly DI, because, according to Aghion et al. (2020) and Henderson and Clark (1990), the necessary research needs to be based on long-term studies, if only to deal with or get around the many pitfalls: unexplored territories, hindrances by researchers occupied by researchers preoccupied with ongoing incremental innovations (Arrow’s replacement effect), pioneering spirit (Arkolakis et al. 2020), welcoming inventors (Bahar et al. 2020), effective will on the part of politicians, “ordinary” conservatism (path dependency syndrome), etc.

As a reminder, Max Weber (2002) wrote the following about the state: “A political enterprise whose administrative leadership successfully claims the monopoly of legitimate physical coercion over a given territory”. Schumpeter (1984), on the other hand, complemented this view, writing that “the two conditions for sustainable growth for a country are the availability of entrepreneurs and administrators and, secondly, the necessary fiscal capacity and savings”. Unlike his predecessors, Keyn developed an economic model centered on the concept of a welfare state that is stimulating, but not lax. His liberalism thus aimed to control and direct economic forces in the interests of justice and social stability (integrated policies of (full) employment and production with flexible wages adapted to demand).

Two important elements can be highlighted here: the simultaneous presence of a strategist state, controlling its spending and the entrepreneurs who support invention. With the military invasion of Ukraine, the Western states did not consider deploying armed forces. On the other hand, with the crisis linked to the virus (and a possible new wave), war on Europe’s doorstep, fears can translate into loss of individual freedom: “Disconnection (cancel culture), framing of thought (boxes and warning messages on Twitter and Facebook) and outright censorship are imposing themselves with great naturalness, as if everything remotely Russian had replaced the non-vaccinated” (Blanchet-Gravel 2022). Even if Germany demonstrated this during the Second World War, is it still possible to envisage open creativity but under strong controls and pressure? “And we go on repeating that times are difficult. But times [...] are always difficult for those who do not love freedom” (Sureau 2022).

It is therefore necessary for there to be "pockets of acceptance" which allow the dissemination and trivialization of new forms of production and manufacturing and, more generally, of societal choices. It would therefore be necessary to change mentalities, at all levels. According to Jancovici (2015), it is an opportunity to be well informed, to understand a critical situation, and to think of being able to participate proactively in a somewhat complicated future. A certain form of "mental plasticity of the social body" risks being exploited, even the opposite, but for the same purpose to be achieved, a "formatting" of people, a displacement by populism of the "social tectonics" (Minc 2005) or a status quo of utilitarian morality, etc. This author also recalls that:

> The population is therefore only fed incorrect information, which is quite annoying since the vote, then the public decision depend on it. Is the criticism of biased information for a public that seeks to have knowledge that is too strongly summarized the cause or the consequence of this state of affairs?

A company's innovation portfolio is selected on the basis of its innovation potential (value for the customer and the company). The evaluation of benefits and risks is particularly difficult for DIs, due to their characteristics such as low comparability with existing technologies and uncertain customer reactions (Weinreich et al. 2020). Figure 2.32 illustrates the presence of a filter informed by expected benefits. Ernst & Young (2021) a priori believe that CEOs would have the impression that institutional investors are rather reluctant to any investment that could harm returns. Their survey shows (see Figure 2.33) that 55% of investors say they think companies should invest in exploring potentially disruptive business models. Note that 67% of investors want companies to invest in DI projects, even if they are risky and may not yield returns in the short term; 73% of investors say that preparing companies for disruption will become a more important factor in investment decisions over the next few years.

However, less than 50% of CEOs surveyed say that their company is not well prepared to take advantage of the opportunities that may emerge from disruptive change.

Company value is not convex and increasing with demand, but is subject to "competitive waves" as rivals enter the market. One company will invest more in innovation if its market is a niche, while another, more historic, company will invest more if its market is broader.

> Positive externalities benefit the incumbent in the development phase, while the innovator has an interest in countering negative externalities (Chevalier-Roignant et al. 2019).

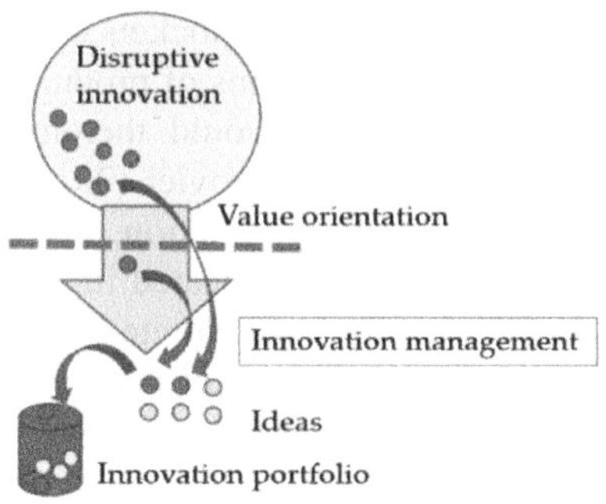

Figure 2.32. *Selection (and storage) of disruptive proposals*

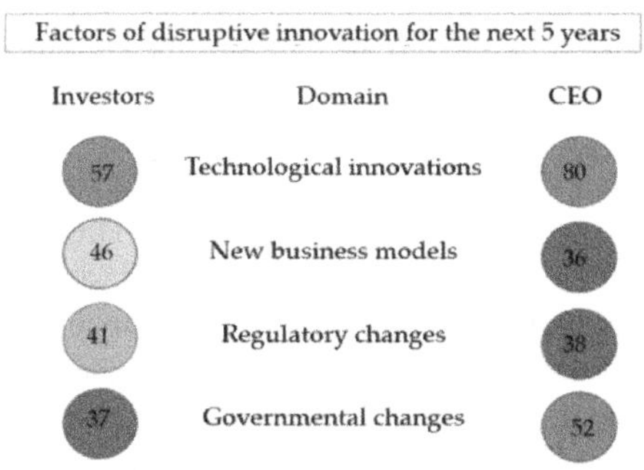

Figure 2.33. *CEOs' position in relation to investors*

2.9. References

Acemoglu, D. and Robinson, J.A. (2012). *Why Nations Fail: The Origins of Power, Prosperity, and Poverty*. Crown, New York.

Aggeri, F. (2020). Vers une innovation responsable. *Esprit*, 3, 40–51.

Aghion, P., Antonin, C., Bunel, S. (2020). *Le pouvoir de la destruction créatrice*. Odile Jacob, Paris.

Aghion, P., Cohen, E., David, B., Gigout-Magiorani, T. (2021). Pour une nouvelle politique industrielle après le Covid-19 [Online]. Available at: https://legrandcontinent.eu/fr/2021/06/04/pour-une-nouvelle-politique-industrielle-apres-le-covid-19/.

Aimé, X. and Arnould, F. (eds) (2021). Qu'est-ce qu'une ontologie computationnelle ? In *Modélisation ontologique et psychologies*. Éditions Matériologiques, Paris.

Akcigit, U., Grisby, J., Nicholas, T. (2017). The rise of American ingenuity: Innovation and innovators of the Golden Age. NBER Working paper no. 23047.

Akrich, M. (1997). The description of technical objects. In *Shaping Technologies/Building Society. Studies in Sociotechnical Change*, Bijker, W.E. and Law, J. (eds). MIT Press, Boston.

Alsaaty, F.M. and Sawyer, G. (2019). Branding entrepreneurial start up disruptive innovation ventures. *Journal of Small Business and Entrepreneurship Development*, 7, 71–81.

Alti Patrimoine (2022). Investissement thématique : le risque d'investir à contre-courant du marché [Online]. Available at: https://www.contrepoints.org/2022/02/26/422362-investissement-thematique-le-risque-dinvestir-a-contre-courant-du-marche.

Alvarez-Pereyre, F. (2003). *L'exigence interdisciplinaire*. MSH, Paris.

Anderson, C. (2012). *Makers : la nouvelle révolution industrielle*. Pearson, Paris.

Anderson, T.R., Daim, T.U., Lavoie, F.F. (2007). Measuring the efficiency of university technology transfer. *Technovation*, 27, 306–318.

André, J.C. (2017a). *From Additive Manufacturing to 3D/4D Printing 1: From Concepts to Achievements*. ISTE Ltd, London, and John Wiley & Sons, New York.

André, J.C. (2017b). *From Additive Manufacturing to 3D/4D Printing 2: Current Techniques, Improvements and their Limitations*. ISTE Ltd, London, and John Wiley & Sons, New York.

André, J.C. (2017c). *From Additive Manufacturing to 3D/4D Printing 3: Breakthrough Innovations: Programmable Material, 4D Printing and Bio-printing*. ISTE Ltd, London, and John Wiley & Sons, New York.

André, J.C. (2019). *Industry 4.0 – Paradoxes and Conflicts*. ISTE Ltd, London, and John Wiley & Sons, New York.

André, J.C. (2020). Engagement solidaire et déconfinement. *Environnement risques et santé*, 19, 238–239.

André, J.C. (2022). *4D Printing 1: Between Disruptive Research and Industrial Applications*. ISTE Ltd, London, and John Wiley & Sons, New York.

Andrew, M., Ford, P.R., Miller, M.T., McRobert, A.P., Foster, N.C., Seerden, G., Littlewood, M.A., Hayes, S.J. (2021). Bridging the gap between science and application: The use of cocreation educational workshops in professional youth soccer. *International Sport Coaching Journal*, 9, 82–99.

Angeler, D., Sundstrom, S., Allen, C. (2016). Deathcore, creativity, and scientific thinking. *Research Ideas and Outcomes*, 2, e8867–6.

Anisov, A.M. (2020). Idea of the proof. *RUDN Journal of Philosophy*, 24, 228–243.

Arbib, J., Dorr, A., Seba, T. (2021). RethinkX – Disruption, implications, and choices [Online]. Available at: https://static1.squarespace.com/static/585c3439be65942f022bbf9b/t/6107fd0ed121a02875c1a99f/1627913876225/Rethinking+Implications.pdf.

Arkolakis, C., Lee, S.K., Peters, M. (2020). European immigrants and the United States' rise to the technical frontiers [Online]. Available at: https://economics.indiana.edu/documents/European-immigrants.pdf.

Arzumanyan, L. and Mayrhofer, U. (2016). L'adoption des outils numériques dans les communautés de pratique. Le cas du Groupe SEB. *Revue française de gestion*, 42, 147–162.

Assemblée Nationale (2022). Proposition de loi pour la création d'un "livret innovation" destiné à renforcer la compétitivité des entreprises selon un fléchage de l'épargne "proposé par Peltier G., Bouley B. le 15/03/2022" [Online]. Available at: https://www.assemblee-nationale.fr/dyn/15/textes/l15b5177_proposition-loi#.

Auernhammer, J. and Roth, B. (2021). The origin and evolution of Stanford University's design thinking: From product design to design thinking in innovation management. *Journal of Product Innovation Management* [Online]. Available at: https://onlinelibrary.wiley.com/doi/full/10.1111/jpim.12594.

Babeau, O. (2022). Développer l'innovation en s'ouvrant sur le monde [Online]. Available at: https://www.lopinion.fr/elections/presidentielle/developper-linnovation-en-souvrant-sur-le-monde-la-tribune-dolivier-babeau.

Bachelard, G. (1963). *Le nouvel esprit scientifique*. PUF, Paris.

Badillo, P.Y. (2013). Les théories de l'innovation revisitées : une lecture communicationnelle et interdisciplinaire de l'innovation ? Du modèle "Émetteur" au modèle communicationnel. *Les Enjeux de l'information et de la communication*, 14, 19–34.

Bahar, D., Choudhury, P., Rapoport, H. (2020). Migrant inventors and the technical advantage of nations. IZA discussion papers no. 12994 [Online]. Available at: https://docs.iza.org/dp12994.pdf.

Baillargeon, N. (2012). Comment "penser" critique ? [Online]. Available at: https://www.acfas.ca/publications/magazine/2012/11/comment-penser-critique.

Balconi, M., Brusoni, S., Oresnigo, L. (2010). In defense of the linear model: An essay. *Research Policy*, 39, 1–13.

Baldwin Partners (2019). Quel est l'intérêt de la méthode C-K à l'heure de l'industrie 4.0 ? [Online]. Available at: https://baldwin-partners.com/industrie-du-futur/quel-est-linteret-de-la-methode-c-k-a-lheure-de-lindustrie-4-0.

Benasayag, M. (1994). *Penser la liberté, la décision, le hasard et la situation*. La Découverte, Paris.

Berger, G. (ed.) (1964). L'idée d'avenir. In *Phénoménologie du temps et prospective*. PUF, Paris.

Bezes, P. (2014). Les Rationalisations des bureaucraties. Perspectives wébériennes sur la nouvelle gestion publique. HDR, Sciences Po, Paris.

Bezes, P. and Musselin, C. (2015). Le new public management : entre rationalisation et marchandisation ? *Une "French touch" dans l'analyse des politiques publiques ?*, Presses de Sciences Po, Paris [Online]. Available at: https://hal-sciencespo.archives-ouvertes.fr/hal-01520666/document.

Bina, O., Inch, A., Pereira, L. (2020). Beyond techno-utopia and its discontents: On the role of utopianism and speculative fiction in shaping alternatives to the smart city imaginary. *Futures*, 115, 102475.

Biron, P. (2022). Recension : effets indésirables des politiques de confinement [Online]. Available at: http://agora.qc.ca/dr-pierre-biron/cui-cui-10-recension-effets-indesirables-des-politiques-de-confinement.

Blanchet-Gravel, J. (2022). Guerre en Ukraine : la peur est là pour rester [Online]. Available at: https://www.contrepoints.org/2022/03/08/422897-guerre-en-ukraine-la-peur-est-la-pour-rester.

Blanco, S. (2008). De quelques signaux faibles à une veille anticipative utile à l'innovation de rupture. *Revue des sciences de gestion*, 231(232), 27–37.

Blankesteijn, M., Bossink, B., van der Sijde, P. (2021). Science-based entrepreneurship education as a means for university-industry technology transfer. *International Entrepreneurship and Management Journal*, 17, 779–808.

Boden, M.A. (2004). *The Creative Mind: Myths and Mechanisms*. Routledge, New York.

Boly, V., Camargo, M., Morel, L. (eds) (2016). Définitions de l'innovation technologique. In *Ingénierie de l'innovation*. Hermès-Lavoisier, Paris.

Bolzani, D., Munari, F., Rasmussen, E., Toschi, L. (2021). Technology transfer offices as providers of science and technology entrepreneurship education. *The Journal of Technology Transfer*, 46, 335–365.

Botsman, R. (2013). The sharing economy lacks a shared definition [Online]. Available at: http://www.fastcoexist.com/3022028/the-sharing-economy-lacks-a-shared-definition.

Bouiss, O. (2020). Qu'est-ce qu'une "crise" ? [Online]. Available at: https://theconversation.com/quest-ce-quune-crise-136026.

Bourdin, S. (2022). Quelle acceptabilité sociale des technologies disruptives ? [Online]. Available at: https://theconversation.com/quelle-acceptabilite-sociale-des-technologies-disruptives-173337.

Bozeman, B., Rimes, H., Youtie, J. (2015). The evolving state-of-the-art in technology transfer research: Revisiting the contingent effectiveness model. *Research Policy*, 44, 34–49.

Busch, L. (2011). *Standards, Recipes for Reality*. MIT Press, Boston.

Bush, V. (1945). Science: The endless frontier [Online]. Available at: https://www.nsf.gov/od/lpa/nsf50/vbush1945.htm.

Buttin, N. and Saffré, B. (2016). *Activer l'économie circulaire*. Eyrolles, Paris.

Caiazza, R. and Volpe, T. (2015). Innovation and its diffusion: Process, actors and actions. *Technology Analysis & Strategic Management*, 29, 181–189.

Canguilhem, G. (2013). *Le normal et le pathologique*. PUF, Paris.

Capaldo, A., Lavie, D., Petruzzelli, A.M. (2014). Knowledge maturity and the scientific value of innovations: The roles of knowledge distance and adoption. *Journal of Management*, 43, 503–533.

Capron Puozzo, I. (2016). Créativité et apprentissage : dilemme et harmonie. *Revue française de pédagogie*, 197, 5–12.

Cargnelli, K. (2020). Private communication.

Casadella, V., Tahi, S., Temple, L. (2017). Créativité et politiques d'innovation. *Innovations*, 53, 5–12.

Casagrande-Seretti, A., Montagna, F., Cascini, G. (2019). A decision support model to assess technological paradigms. *International Journal of Technology Management*, 80, 61–84.

Castoriadis, C. (1999). *Figures du pensable*. Le Seuil, Paris.

Chardel, P.A. and Dartiguepeyrou, C. (2018). Être, temps et différences : pour une approche différencialiste du temps à l'ère numérique. In *@ la recherche du temps. Individus hyperconnectés, société accélérée. Tensions et transformations*, Aubert, N. (ed.). Eres, Paris.

Charue-Duboc, F. and Gastaldi, L. (2017). Le pilotage des processus d'innovation amont. Vers de nouvelles modalités de couplage entre technologies et usages. *Revue française de gestion*, 43, 23–42.

Charue-Duboc, F., Gastaldi, L., Bertin, E. (2020). Des communautés d'experts internes comme facilitateur de l'innovation. *Revue française de gestion*, 287, 81–98.

Chaurand, N. and Brauer, M. (2008). La déviance. *Revue électronique de psychologie sociale*, 3, 9–23.

Chemin-Bouzir, C. (2019). L'imaginaire : leurre générateur d'agressivité ou consistance génératrice de dynamiques sociales créatives ? Le cas de Conseils départementaux et d'associations d'aide à domicile. *Revue internationale de psychosociologie et de gestion des comportements organisationnels*, 60, 63–80.

Chevalier-Roignant, B., Flath, C.M., Trigeorgis, L. (2019). Disruptive innovation and market entry and production flexibility in heterogeneous oligopoly. *Production and Operations Management*, 28, 1641–1657.

Chouteau, M., Forest, J., Nguyen, C. (2020). *Innovation for Society: The P.S.I. Approach*. ISTE Ltd, London, and John Wiley & Sons, New York.

Christensen, C.M. (1997). *The Innovator's Dilemma: When New Technologies Cause Great Firms to Fail*. Harvard Business School Press, Brighton.

Cina, M. (2021). L'imaginaire selon Castoriadis pour comprendre la créativité dans les organisations. *Innovations*, 66, 17–45.

Clarysse, B., Tartari, V., Salter, A. (2011). The impact of entrepreneurial capacity, experience and organizational support on academic entrepreneurship. *Research Policy*, 40, 1084–1093.

CNRS (2022). La recherche, l'avenir de l'Europe [Online]. Available at: https://www.cnrs.fr/fr/cnrsinfo/la-recherche-lavenir-de-leurope?sstc=u8216nl6387.

Cohen, D. (2009). *La prospérité du vice – Une introduction inquiète à l'économie*. Albin Michel, Paris.

Cohen, D. and Feld, B. (2019). *Do More Faster: Techstars Lessons to Accelerate Your Startup*. John Wiley & Sons, New York.

Cornu, J.M. (2008). *Nouvelles technologies, nouvelles pensées ? La convergence des NBIC*. Fyp, Limoges.

D'Espagnat, B. (1983). *In Search of Reality*. Springer, Berlin.

Daloz, P., Johnson, P., Massart, S., Le Masson, P., Weil, B. (2020). Shaping the unknown with virtual universes. The new fuel for innovation. Global Innovation Index, Cornell University, INSEAD, WIPO [Online]. Available at: https://hal-mines-paristech.archives-ouvertes.fr/hal-03042503/document.

De Cock, C., Rehn, A., Berry, D. (2013). For a critical creativity: The radical imagination of Cornelius Castoriadis. In *Handbook of Research on Creativity*, Thomas, K. and Chan, J. (eds). Edward Elgar, Cheltenham.

De Fillippi, R., Grabher, G., Jones, C. (2007). Introduction to paradoxes of creativity: Managerial and organizational challenges in the cultural economy. *Journal of Organizational Behavior*, 28, 511–521.

De Geuser, F. and Fiol, M. (2003). Faire face aux situations complexes : la blessure narcissique des managers. In *Peut-on former les dirigeants ? L'apport de la recherche*, Moingeon, B. (ed.). L'Harmattan, Paris.

De Grave, A. (2015). La fin des Léviathans. In *Société collaboratrice – La fin des hiérarchies*, Fillipova, D. (ed.). Rue de l'échiquirer, Paris.

Deloitte (2021). Smart factory transformation: The time is now [Online]. Available at: https://www.forbes.com/sites/deloitte/2021/01/21/smart-factory-transformation-the-time-is-now/?sh=112a67bf6470.

Deming, P. and Dunham, R. (2010). *The Innovator's Way: Essential Practices for Successful Innovation*. MIT Press, Boston.

Deserti, A., Real, M., Schmittinger, F. (2022). *Co-creation for Responsible Research and Innovation*. Springer, Berlin.

Desgagné, S. and Bednarz, N. (2005). Médiation entre recherche et pratique en éducation : faire de la recherche "avec" plutôt que "sur" les praticiens. *Revue des sciences de l'éducation*, 31, 245–258.

Dodet, M. (2001). Innovation et société. *Passages*, 106(107), 22–23.

Dosi, G. (1982). Technological paradigms and technological trajectories: A suggested interpretation of the determinant and direction of technological change. *Research Policy*, 11, 147–162.

Dosi, G. (1988). Sources, procedures and micro-economic effects of innovation. *Journal of Economic Literature*, 26, 1120–1171.

Dufresne, J. (2022). La liberté, le mal, la vie, la mort [Online]. Available at: http://agora.qc.ca/jacques-dufresne/la-liberte-le-mal-la-vie-la-mort.

Dupuy, J.P. (2018). Nous n'avons pas peur de ce qui devrait nous terrifier. In *De quoi avons-nous peur*, Birnbaum, J. (ed.). Folios-Essais, Paris.

Egger, P. and Loumeau, N. (2019). The economic geography of innovation [Online]. Available at: https://voxeu.org/article/economic-geography-innovation.

Einstein, A. and Infeld, L. (1938). *The Evolution of Physics*. Cambridge University Press, Cambridge.

Ellul, J. (1954). *La Technique ou l'enjeu du siècle*. Armand Colin, Paris.

Ernst and Young (2021). How can you be both the disruptor and the disrupted? [Online]. Available at: https://assets.ey.com/content/dam/ey-sites/ey-com/en_gl/topics/digital/ey-how-can-you-be-both-the-disruptor-and-the-disrupted.pdf.

European Commission (1995). Livre vert sur l'innovation [Online]. Available at: https://europa.eu/documents/comm/green_papers/pdf/com95_688_fr.pdf.

European Commission (2009). Année européenne de la créativité et de l'innovation [Online]. Available at: https://eur-lex.europa.eu/legal-content/FR/TXT/?uri=LEGISSUM%3Al29020.

European Commission (2021). La Commission lance le Conseil européen de l'innovation pour contribuer à transformer des idées scientifiques en innovations prometteuses [Online]. Available at: https://ec.europa.eu/commission/presscorner/detail/fr/IP_21_1185.

Eschenfelder, K.R. and Shankar, K. (2017). Organizational resilience in data archives: Three case studies in social science data archives. *Data Science Journal*, 16, 1–12.

Escudié, M.P. (2013). Gaston Berger, les sciences humaines et les sciences de l'ingénieur. Un projet de réforme de la société. PhD Thesis, Université de Lyon II, Lyon [Online]. Available at: http://institut-gaston-berger.insa-lyon.fr/sites/institut-gaston-berger.insa-lyon.fr/files/doctorat_mpescudie.pdf.

Etzkowitz, H. and Leydesdorff, L. (2000). The dynamics of innovation: From national systems and "mode 2" to a triple helix of university–industry–government relations. *Research Policy*, 29, 109–123.

Fablab (2016). Qu'est-ce qu'un Fablab ? [Online]. Available at: http://carrefour-numerique.cite-sciences.fr/fablab/wiki/doku.php?id=charte.

Fanvoice (2021). Synthèse de campagne et si la co-innovation servait la science ? [Online]. Available at: https://cdn.futura-sciences.com/sources/mkt/sondage-futura-coinnov.pdf.

Fernandez, A. (2001). *Le bon usage des technologies expliqué au manager*. Eyrolles, Paris.

Fœssel, M. (2015). Une ambition philosophique par gros temps. *Esprit*, 8(9), 6–11.

Follett, M.P. (1949). *Freedom and Coordination. Lectures in Business Organization Management*. Publication Trust Limited, London.

Fombelle, P.W., Voorhees, C.M., Jenkins, M.R., Sidaoui, K., Benoit, S., Gruber, T., Gustafsson, A., Abosag, I. (2020). Customer deviance: A framework, prevention strategies, and opportunities for future research. *Journal of Business Research*, 116, 387–400.

Foray, D. (1992). Choix de la technique, rendements croissants et processus historiques. In *La technoscience. Les fractures du discours*, Pradès, J. (ed.). L'Harmattan, Paris.

Foray, D. (2002). Ce que l'économie néglige ou ignore en matière d'analyse de l'innovation. In *Les logiques de l'innovation*, Adler, N. (ed.). La Découverte, Paris.

Forest, J., Chouteau, M., Nguyen, C. (2011). Conceptions de l'innovation et formations de l'ingénieur. *Les cahiers du musée des confluences*, 7, 37–47.

France Stratégie (2022a). Programme de travail [Online]. Available at: https://www.strategie.gouv.fr/sites/strategie.gouv.fr/files/atoms/files/fs-2022-programme_de_travail_2022_-_mars.pdf.

France Stratégie (2022b). Les Métiers en 2030. Quels métiers recruteront à l'horizon 2030 ? [Online]. Available at: https://www.strategie.gouv.fr/sites/strategie.gouv.fr/files/atoms/files/fs-2022-pmq-dp-metiers_2030_quels_metiers_recruteront_a_lhorizon_2030-mars.pdf.

Freeman, C. (1988). À quoi tiennent la réussite ou l'échec des innovations dans l'industrie ? Culture technique [Online]. Available at: http://documents.irevues.inist.fr/bitstream/handle/2042/31862/C&T_1988_18_30.pdf?sequence=1.

Freeman, C., Clark, J., Soete, L. (1982). *Unemployment and Technical Innovation: A Study of Long Waves and Economic Development*. Greenwood Press, Westport.

Frese, M., Bausch, A., Schmidt, P., Rauch, A., Kabst, R. (2012). Evidence-based entrepreneurship: Cumulative science, action principles and bridging the gap between science and practice. *Foundations and Trends in Entrepreneurship*, 8, 1–62.

Friedman, G. (1952). Les conséquences sociales du progrès technique. *UNESCO. Bulletin International des Sciences Sociales*, 4, 251–272.

Frochot, C., Schaer, E., André, J.C. (2020). Il ne faut pas avoir peur des "nano", c'est juste une nano-crise mal anticipée. *Environnement risques et santé*, 19, 446–466.

Gaines, B.R. (2019). From facilitating interactivity to managing hyper-connectivity: 50 years of human–computer studies. *International Journal of Human-Computer Studies*, 131, 4–22.

Gardey, D. (2008). *Écrire, calculer, classer. Comment une révolution de papier a transformé les sociétés contemporaines (1800–1940)*. La Découverte, Paris.

Garel, G. and Rosier, R. (2007). Gérer l'exploration : le cas des technologies à haut potentiel. In *Congrès annuel de l'Association des sciences administratives du Canada*, Ottawa.

Gaudin, B. (2013). Invention, diffusion and linear models of innovation. Project on the intellectual history of innovation [Online]. Available at: http://www.csiic.ca/PDF/AnthropologyPaper15.pdf.

GEAB 161 (2022). Panorama 2022 : tendances et anticipations [Online]. Available at: https://geab.eu/panorama-2022-tendances-et-anticipations/?utm_source=sendinblue&utm_campaign=Le%20GEAB%20de%20janvier%20est%20disponible%20Panorama%202022%20-%20Tendances%20et%20anticipations&utm_medium=email.

Gerschel, E., Martinez, A., Mejean, I. (2020). Propagation des chocs dans les chaînes de valeur internationales : le cas du coronavirus. Note IPP no. 53 [Online]. Available at: https://www.ipp.eu/news/note-ipp-n53-propagation-des-chocs-dans-les-chaines-de-valeur-internationales-le-cas-du-coronavirus/.

Gershenfeld, N. (2012). How to make almost anything: The digital fabrication revolution. *Foreign Affairs*, 91, 43–57.

Geslin, P. (2017). *Inside Anthropotechnology: User and Culture Centered Experience*. ISTE Ltd, London, and John Wiley & Sons, New York.

Giget, M. (2013). Nous vivons une période de poussée technologique considérable [Online]. Available at: http://www.01net.com/actualites/marc-giget-nous-vivons-une-periode-de-poussee-technologique-considerable-600167.html.

Godin, B. (2006). The linear model of innovation: The historical construction of an analytical framework. *Science, Technology, and Human Values*, 31, 639–667.

Goglio-Primard, K. and Soulier, E. (2018). Connaissance et technologie dans les communautés d'innovation. *Systèmes d'information et management*, 23, 3–9.

Gold, E.R. and Morgan, M. (2019). For innovation, open science means open for business [Online]. Available at: https://www.theglobeandmail.com/business/commentary/article-for-innovation-open-science-means-open-for-business/.

Goldberg, A. (2022). Associate professor of organizational behavior [Online]. Available at: https://www.gsb.stanford.edu/faculty-research/faculty/amir-goldberg.

GoTronic (2016). Carte des Fablabs [Online]. Available at: http://www.gotronic.fr/ins-carte-des-fablabs-50.htm.

Granger, G.G. (1992). À quoi sert l'épistémologie ? *Droit et société*, 20(21), 39–44.

Granstrand, O. and Holgersson, M. (2020). Innovation ecosystems: A conceptual review and a new definition. *Technovation*, 102098, 90–91.

Groves, C. (2017). Emptying the future: On the environmental politics of anticipation. *Futures*, 92, 29–38.

Guilford, J.P. (1950). Creativity. *American Psychologist*, 5, 444–454.

Guilford, J.P. (1959). Traits of creativity. In *Creativity and Its Cultivation*, Anderson, H.H. (ed.). Harper Collins, New York.

Guillaume, L.P., Delmas, C., Goglio-Primard, K. (2017). Schneider Electric pilote les communautés de pratique avec le programme Communities@Work. In *Les communautés d'innovation : de la liberté créatrice à l'innovation organisée*, Sarazin, B., Cohendet, P., Simon, L. (eds). Éditions EMS, Caen.

Harroche, A. (2021). Gouverner par les inégalités : la mise en œuvre d'une initiative d'excellence dans l'enseignement supérieur et la recherche. PhD Thesis, Institut d'études politiques de Paris, Sciences-Po, Paris.

Harryson, S., Kliknaite, S., Dudkowski, R. (2008). Flexibility in innovation through external learning: Exploring two models for enhanced industry–university collaboration. *International Journal of Technology Management*, 41, 109–137.

Haudeville, B. and Le Bas, C. (2016). L'innovation frugale, paradigme technologique naissant ou nouveau modèle d'innovation ? *Innovations*, 51, 9–25.

Hayter, C.S., Lubynsky, R., Maroulis, S. (2016). Who is the academic entrepreneur? The role of graduate students in the development of university spinoffs. *The Journal of Technology Transfer*, 41, 1–18.

Heracleous, L. and Robson, D. (2021). Organizations tend to see rebels as troublemakers – But suppressing these individuals and their ideas could backfire [Online]. Available at: https://www.bbc.com/worklife/article/20210528-positive-deviants-why-rebellious-workers-spark-gr-ideas.

Hoffmann, S. (1963). The French political community. In *Search of France: The Economy, Society, and Political System in the Twentieth Century*, Hoffmann, S. (ed.). Harvard University Press, Brighton.

Hohberger, J. (2016). Diffusion of science-based inventions. *Technological Forecasting and Social Change*, 104, 66–77.

Holford, W.D. (2019). The future of human creative knowledge work within the digital economy. *Futures*, 105, 143–154.

Howells, J. (2006). Intermediation and the role of intermediaries in innovation. *Research Policy*, 35, 715–728.

Institut Montaigne (2021). Innovation française : nos incroyables talents [Online]. Available at: https://www.institutmontaigne.org/ressources/pdfs/publications/innovation-francaise-nos-incroyables-talents-rapport.pdf.

Jakobsen, S.E., Fløysand, A., Overton, J. (2019). Expanding the field of Responsible Research and Innovation (RRI) – From responsible research to responsible innovation. *European Planning Studies*, 27, 2329–2343.

Jancovici, J.M. (2015). *Dormez tranquille jusqu'en 2100 et autres malentendus sur le climat et l'énergie*. Odile Jacob, Paris.

Jarroson, B. (2017). *Oser la confiance : propos sur l'engagement des dirigeants*. Eyrolles, Paris.

Johansson-Sköldberg, U., Woodilla, J., Çetinkaya, M. (2013). Design thinking: Past, present and possible futures. *Creativity and Innovation Management*, 22, 121–146.

Jonas, H. (2013). *Essais philosophiques. Changement et permanence, sur la possibilité de comprendre l'histoire*. Vrin, Paris.

Kant, I. (1987). *Critique of Judgement*. Hackett Publishing Company, Vrin, Paris.

Kelly, B., Papanikolaou, D., Seru, A., Taddy, M. (2021). Measuring technological innovation over the long run. *American Economic Review: Insights*, 3, 303–320.

Kennedy, J.F. (2005). *80 Hommes pour changer le monde : entreprendre pour la planète*. Le Livre de poche, Paris.

Keynes, J.M. (2016). *The General Theory of Employment, Interest, and Money*. Wordsworth Editions, London.

Keynes, J.M. (2021). *La vertu et le bonheur suivi de possibilités économiques pour nos petits-enfants*. Les Belles Lettres, Paris.

Kijkuit, B. and van den Ende, J. (2007). The organizational life of an idea: Integrating social network, creativity and decision-making perspectives. *Journal of Management Studies*, 44, 863–882.

King, A. (1999). *Géopolitique du chaos*. Folio Actuel, Paris.

Kirzner, I.M. (1979). *Perception, Opportunity and Profit: Studies in the Theory of Entrepreneurship*. Chicago University Press, Chicago.

Koenig, G. and Meyer N.M.P. (2022). Entretien avec Gaspard Koenig : simplifions-nous la vie ! [Online]. Available at: https://www.contrepoints.org/2022/03/20/423679-entretien-avec-gaspard-koenig-simplifions-nous-la-vie.

Kollmann, T., Stöckmann, C., Hensellek, S., Kensbock, J. (2016). European startup monitor 2016 [Online]. Available at: http://europeanstartupmonitor.com/esm/esm-2016/.

Komporozos-Athanasiou, A. and Fotaki, M. (2015). A theory of imagination for organization studies using the work of Cornelius Castoriadis. *Organization Studies*, 36, 321–342.

Kotze, P. and Purgathofer, P. (2007). A framework for design exercises – From creativity to real-world use [Online]. Available at: https://publik.tuwien.ac.at/files/pub-inf_5251.pdf.

Kraujalienė, L. (2019). Comparative analysis of multicriteria decision-making methods evaluating the efficiency of technology transfer. *Business, Management and Education*, 17, 72–93.

Kunegel, J.R. (2019). L'influence de l'incubateur sur les missions du bureau de transfert de technologie au sein des universités : le cas de l'University City Science Center (États-Unis). *Marché et organisations*, 34, 109–131.

Lampel, J., Lant, T.K., Shamsie, J. (2000). Balancing act: Learning form organizational practices in cultural industries. *Organization Science*, 11, 263–269.

Laperche, B. (2020). Mutation des trajectoires des entreprises et innovation : capital savoir et opportunités financières. *Marché et organisations*, 39, 91–110.

Latour, B. (1984). *Les microbes. Guerre et paix*. Métailié, Paris.

Lavis, A. (2021). *L'Imprévu. Que faire lorsqu'on ne sait plus ?* Autrement, Paris.

Le Du, L. (2017). Modéliser l'expansion des imaginaires en conception : dynamique des imaginaires, ingénierie de stimulation et nouvelles organisations de l'innovation. PhD Thesis, Université Paris Sciences et Lettres, Paris.

Le Masson, P., Weil, B., Hatchuel, A. (2006). *Les processus d'innovation, conception innovante et croissance des entreprises*. Hermès-Lavoisier, Paris.

Le Masson, P., Hatchuel, A., Weil, B. (2018). Théorie C-K. Fondements et implications d'une théorie de la conception. *Techniques de l'ingénieur*, J8115 v1.

Le Pogam, Y. (1997). Corporéisme et individualisme hédoniste. *Corps et culture*, 2 [Online]. Available at: http://corpsetculture.revues.org/388.

Léac, J.P. (2022). Qu'est-ce-que le design thinking ? [Online]. Available at: https://www.lescahiersdelinnovation.com/qu-est-ce-que-le-design-thinking/.

Leclair, M. (2018). Trouble créatif et position évasive, pratiques de créatifs en contexte marchand. *Management International*, 22, 73–86.

Leef, G. (2022). Thomas Piketty récidive [Online]. Available at: https://www.contrepoints.org/2022/02/01/420655-thomas-piketty-recidive.

Lesourne, J. and Randet, D. (2016). *La recherche et l'innovation en France*. Odile Jacob, Paris.

Lewis, W.A. (1980). The slowing down of the engine of growth. *The American Economic Review*, 70, 555–564.

Lhoste, E.F. (2017). Les fablabs transforment-ils les pratiques de médiation ? *Cahiers de l'action*, 48, 15–22.

Link, A.N. and Scott, J.T. (2020). Distance between scientific knowledge and applications. *Diffusion of Science-Based Invention. Economics of Innovation and New Technology*, 29, 489–500.

Love, J.H. and Roper, S. (2009). Organizing innovation: Complementarities between cross-functional teams. *Technovation*, 29, 192–203.

Luzi, J. (2014). Du catastrophisme lucratif au totalitarisme vert ? Une lecture de 1984, de George Orwell. *Écologie & politique*, 48, 125–141.

Malecki, E.J. (2010). Global knowledge and creativity: New challenges for firms and regions. *Regional Studies*, 44, 1033–1052.

Manuel d'Oslo (1997). *Principes directeurs proposés pour le recueil et l'interprétation des données sur l'innovation technologique*. OECD, Paris.

Maslow, A. (1954). *Motivation and Personality*. Harper Collins, New York.

Maslow, A. (1966). *The Psychology of Science*. Harper Collins, New York.

Mathias, D.R.F. (2019). Investigating and characterizing the coupling of LEGO and 3D printing. PhD Thesis, University of Bristol, Bristol [Online]. Available at: https://research-information.bris.ac.uk/ws/portalfiles/portal/225056783/DM_Thesis_Corrected.pdf.

Matt, M. and Schaeffer, V. (2018). Building entrepreneurial ecosystems conducive to student entrepreneurship: New challenges for universities. *Journal of Innovation Economics & Management*, 1, 9–32.

Mavri, M. (2015). Redesigning a production chain based on 3D printing technology. *Knowledge & Process Management*, 22, 141–147.

McCown, R.L. (2001). Learning to bridge the gap between science-based decision support and the practice of farming: Evolution in paradigms of model-based research and intervention from design to dialogue. *Australian Journal of Agriculture Research*, 52, 549–571.

McKinsey (2022). The data-driven enterprise of 2025 [Online]. Available at: https://www.mckinsey.com/business-functions/mckinsey-analytics/our-insights/the-data-driven-enterprise-of-2025?cid=other-eml-dre-mip-mck&hlkid=a774e1a8976d4a078406df70a8406fd5&hctky=11339392&hdpid=99381738-0cb3-4708-903e-1ba5579af7aa.

Merle, B. (2019). Fonder une start-up, une aventure à risques mais qui peut rapporter gros ! [Online]. Available at: https://www.futura-sciences.com/tech/questions-reponses/tech-fonder-start-up-aventure-risques-mais-peut-rapporter-gros-12444/.

Meyer, J.W. (1994). Rationalized environments. In *Institutional Environments and Organizations: Structural Complexity and Individualism*, Scott, R. and Meyer, J.W. (eds). Sage, London.

Meyer, J.W. and Rowan, B. (1977). Institutionalized organizations: Formal structure as myth and ceremony. *American Journal of Sociology*, 83, 340–363.

Meyer, J.W., Boli, J., Thomas, G.M. (1987). Ontology and rationalization in the western cultural account. In *Institutional Structure: Constructing State, Society, and the Individual*, Thomas, G.M., Meyer, J.W., Ramirez, F.O., Boli, J. (eds). Sage, London.

Meyer, J.W., Drori, G.S., Hwang, H. (2006). World society and the proliferation of formal organizations. In *Globalization and Organization – World Society and Organizational Change*, Drori, G.S., Meyer, J.W., Hwang, H. (eds). Oxford University Press, Oxford.

Micheli, P., Wilner, S.J.S., Bhatti, S.H., Mura, M., Beverland, M.B. (2021). Doing design thinking: Conceptual review, synthesis, and research agenda. *Journal of Product Innovation Management*. doi: 10.1111/jpim.12466.

Minc, A. (2004). *Ce monde qui vient*. Grasset, Paris.

MITI-CNRS (2022). Soutenir les prises de risques, la rupture et l'émergence [Online]. Available at: https://miti.cnrs.fr/.

MKGI (2019). Innovation in Europe. Changing the game to regain a competitive edge [Online]. Available at: https://www.mckinsey.com/~/media/McKinsey/Featured%20Insights/Innovation/Reviving%20innovation%20in%20Europe/MGI-Innovation-in-Europe-Discussion-paper-Oct 2019-vF.ashx.

Morace, C. and Coadour, D. (2018). Former des ingénieurs au transfert de technologie : entre dimensions professionnelles et dimensions culturelles. In *Savoirs "Formation et professionnalisation des ingénieurs : articulations et tensions"*. L'Harmattan, Paris.

Morel, C. (2002). *Les décisions absurdes*. Gallimard, Paris.

Moussion, A. (2015). Impression 3D : quel impact sur la propriété intellectuelle ? L'INPI répond dans une interview [Online]. Available at: http://www.priximprimante3d.com/loi/.

NAE Perspectives (2022). Technoscientific research: A missing term in R&D discourse [Online]. Available at: https://www.nae.edu/242564/NAE-Perspectives.

NAP (2015). Healthy, resilient, and sustainable communities after disasters: Strategies, opportunities, and planning for recovery [Online]. Available at: http://nap.edu/18996.

Nature Index (2021). 2021 tables: Countries/territories [Online]. Available at: https://www.natureindex.com/annual-tables/2021/country/all.

Naudon, F. (2013). Comment le profane joue en faveur du décloisonnement. *Hermès – La Revue. Interdisciplinarité : entre disciplines et indiscipline*, Paris.

Nazin, R. (2016). Knowledge integration with specificity preservation [Online]. Available at: https://hal.inria.fr/hal-01718230/file/Article_NAZIN_Remi.pdf.

Nesta (2021). A compendium of innovation methods [Online]. Available at: https://media.nesta.org.uk/documents/Compendium-of-Innovation-Methods.pdf?_ga=2.216779926.1045554066.1553702423-715356215.1553702423.

Nicolescu, B. (1996). *La transdisciplinarité – manifeste*. Éditions du Rocher, Monaco.

Nietzsche, F. (1979). *Fragments posthumes 1886–1887*. NRF, Paris.

Noailles, P. (2011). De l'innovation à l'innovateur. Pour une approche structuraliste de l'innovation. *Revue des sciences de gestion*, 13(28), 247–248.

Norman, D.A. and Verganti, R. (2014). Incremental and radical innovation: Design research vs. technology and meaning change. *Design Issues*, 30, 78–96.

O'Kane, C. (2018). Technology transfer executives' backwards integration: An examination of interactions between university technology transfer executives and principal investigators. *Technovation*, 64(77), 76–77.

O'Reilly, C.A. and Tushman, M.L. (2004). The ambidextrous organization. *Harvard Business Review*, 82, 74–81.

OECD (2010). OECD innovation strategy [Online]. Available at: https://www.oecd.org/site/innovationstrategy/.

OECD (2013). NNI. Synthesis report [Online]. Available at: http://eprints.internano.org/1918/1/Washington%20Symposium%20Report_final.pdf.

OECD (2021a). The design and implementation of mission-oriented innovation policies – A new systemic policy approach to address societal challenges [Online]. Available at: https://www.oecd-ilibrary.org/docserver/3f6c76a4-en.pdf?expires=1648450326&id=id&accname=guest&checksum=3FF5455B0F0D5647E050D98D5CFA17B1.

OECD (2021b). Knowledge co-creation in the 21st century – A cross-country experience-based policy report [Online]. Available at: https://www.oecd-ilibrary.org/docserver/c067606f-en.pdf?expires=1648400248&id=id&accname=guest&checksum=8354A406FF9D9524F598FBA0DF6EE5D0.

Orfali, B. (2005). *La société face aux événements extraordinaires*. Zagros, Paris.

Orwell, G. (2000). *Animal Farm*. Penguin Books, London.

Palmberg, C., Pajarinen, M., Nikulainen, T. (2010). Transferring science-based technologies to industry. Does nanotechnology make a difference? *Technovation*, 30, 3–11.

Pépin, A., Giroux, P., Monney, N. (2019). Qu'est-ce qu'un laboratoire créatif ? [Online]. Available at: http://rire.ctreq.qc.ca/2019/12/laboratoire-creatif/.

Pessoa, F. (1992). *Le Livre de l'intranquillité*. Christian Bourgeois, Paris.

Petropoulos, G. (2017). An economic review of the collaborative economy. Bruegel Policy Contributions [Online]. Available at: https://www.econstor.eu/handle/10419/173101.

Plagnol, A. (2020). Confinement mental et contention. *PSN*, 18, 9–15.

Popper, K. (2002). *The Logic of Scientific Discovery*. Routledge, London and New York.

Poszytek, P. (2021). The landscape of scientific discussions on the competencies 4.0 concept in the context of the 4th industrial revolution. A bibliometric review. *Sustainability (Basel, Switzerland)*, 13, 6709.

Puech, M. (2016). *Homo Spiens Technologicus*. Le Pommier, Paris.

Rau, H., Goggins, G., Fahy, F. (2018). From invisibility to impact: Recognizing the scientific and societal relevance of interdisciplinary sustainability research. *Research Policy*, 47, 266–276.

Raynaud, D. (2021). *Sociologie fondamentale. Étude d'épistémologie*. Éditions Matériologiques, Paris.

Recker, J.V. (2016). What triggers innovation diffusion? Intermediary organizations and geography in cultural and science-based industries. *Environment and Planning C: Government and Policy*, 34, 1058–1075.

Reiter-Palmon, R. and Illies, J.J. (2004). Leadership and creativity: Understanding leadership from a creative problem solving perspective. *The Leadership Quarterly*, 15, 55–77.

Rentzos, L., Doukas, M., Mavrikios, D., Mourtzis, D., Chryssolouris, G. (2014). Integrating manufacturing education with industrial practice using teaching factory paradigm: A construction equipment application. *Procedia CIRP*, 17, 189–194.

Reversat, N. and André, J.C. (2022). *Covid 19 – À qui, ou à quoi se fier ?* Awaiting publication.

Rias, A.L. (2017). Créativité par et pour la fabrication additive : proposition d'une méthodologie outillée. PhD Thesis, ENSAM, Paris.

Richtner, A. and Löfsten, H. (2014). Managing in turbulence: How the capacity for resilience influences creativity. *R&D Management*, 44, 137–151.

Rigo, D. (2021). Global value chains and technology transfer: New evidence from developing countries. *Review of World Economics*, 157, 271–294.

Rodrik, D. (2007). *One Economics, Many Recipes: Globalization, Institutions, and Economic Growth*. Princeton University Press, Princeton.

Roquette, M.L. (2007). *La créativité*. PUF, Paris.

Rosa, H. (2018). *Remède à l'accélération*. Champs Essais, Paris.

Rothwell, R. (1992). Successful industrial innovation: Critical factors for the 1990s. *R&D Management*, 22, 221–239.

Rothwell, R. (1994). Industrial innovation: Success, strategy, trends. In *The Handbook of Industrial Innovation*, Dodgson, M. and Rothwell, R. (eds). Edward Elgar, Aldershot.

Rothwell, R. and Zegveld, W. (1982). *Innovation and the Small and Medium Sized Firm: Their Role in Employment and in Economic Change*. Frances Pinter, London.

Ruano-Borbalan, J.C. (2017a). La "grande standardisation" et les transformations du savoir dans les sociétés technoscientifiques. *Innovations*, 52, 17–41.

Ruano-Borbalan, J.C. (2017b). Technosciences en société : les voies multiples de la légitimation des savoirs. *Innovations*, 52, 5–15.

Ruwet, C. (2017). Pour en finir avec la confusion entre normes et standards. Quelques repères pour situer la normalisation dans l'espace-temps normatif. In *Puissances de la norme*, Le Goff, J. and Onnée, S. (eds). EMS Éditions, Paris.

Rylander, E.A., Navarro, A.U., Amacker, A. (2021). Design thinking as sensemaking. Developing a pragmatist theory of practice to (re)introduce sensibility. *Journal of Product Innovation Management* [Online]. Available at: https://onlinelibrary.wiley.com/doi/full/10.1111/jpim.12604.

Sadin, E. (2021). *La silicolonisation du monde*. Éditions de l'échappée, Paris.

de Saint-Simon, C.H. (2012). *La parabole & sur la querelle des abeilles et des frelons*. Ores et Déjà, Paris.

Salgian, A., Nakra, T.M., Ault, C., Wang, Y. (2013). Conducting robots. Bridging the gap between science, technology and the arts in the undergraduate curriculum. In *IEEE Integrated STEM Education Conference*. Princeton University Press, Princeton.

Santos Gomes, M.A. and Kovaleski, J.L. (2019). International technology transfer and the irruptive processes: An analysis model for the offset policy. *International Journal of Technology Transfer and Commercialization*, 16, 103–117.

Saravi, M. and Kordlouie, H. (2009). The role of governmental organizations services in public trust development. In *Managing Services Across Continents*, Altinay, L., Bowen, D., Ekinci, Y. (eds). Oxford Brookes University Business School, Oxford.

Schaeffer, V. (2019). L'université entrepreneuriale : éléments historiques et débats. *Marché et organisations*, 34, 87–108.

Schaeffer, V., Öcalan-Özel, S., Pénin, J. (2018). The complementarities between formal and informal channels of university–industry knowledge transfer: A longitudinal approach. *The Journal of Technology Transfer*, 45, 1–25.

Schmookler, J. (1966). *Invention and Economic Growth*. Harvard University Press, Boston.

Schneider, C. and Lösch, A. (2019). Visions in assemblages: Future-making and governance in FabLab. *Futures: The Journal of Policy, Planning and Futures Studies*, 109, 203–212.

Schopenhauer, A. (1997). *De la quadruple racine du principe de raison suffisante*. Vrin, Paris.

Schumpeter, J.A. (1983). *Capitalisme, socialisme et démocratie*. Payot, Paris.

Schumpeter, J.A. (ed.) (1984). La crise de l'Etat fiscal. In *Impérialisme et classes sociales*. Flammarion, Paris.

Schwartz, B. (2021). La victoire posthume de Bertrand Schwartz [Online]. Available at: https://www.telos-eu.com/fr/societe/la-victoire-posthume-de-bertrand-schwartz.html.

Servier, J. (1991). *Histoire de l'utopie*. Folio-Essais, Paris.

Shah, S.K. and Panhke, E.C. (2014). Parting the ivory curtain: Understanding how universities support a diverse set of startups. *The Journal of Technology Transfer*, 39, 780–792.

Shmeleva, N., Gamidullaeva, L., Tolstykh, T., Lazarenko, D. (2021). Challenges and opportunities for technology transfer networks in the context of open innovation: Russian experience. *Journal of Open Innovation: Technology, Market and Complexity*, 7, 197.

Siegel, D.S. and Wright, M. (2015). Academic entrepreneurship: Time for a rethink? *British Journal of Management*, 26, 582–595.

Simmel, G. (1999). *Sociologie. Études sur les formes de la socialisation*. Presses universitaires de France, Paris.

Simon, H.A. (1976). From substantive to procedural rationality. In *Methods and Appraisal in Economics*, Latsis, S.J. (ed.). Cambridge University Press, Cambridge.

Sloterdijk, P. (2000). *Critique de la raison cynique*. C. Bourgeois, Paris.

Sloterdijk, P. (2006). *Le palais de cristal : à l'intérieur du capitalisme planétaire*. Moren Sell, Paris.

Snider, C., Dekoninck, E., Culley, S. (2016). Beyond the concept: Characterisations of later-stage creative behaviour in design. *Research in Engineering Design*, 27, 265–289.

Spencer, H. (2017). *First Principles*. John Childs and Son, London [Online]. Available at: https://www.gutenberg.org/cache/epub/55046/pg55046-images.html.

Stacey, K., Fontanella-Khan, J., Palma, S. (2021). Big tech companies snap up smaller rivals at record pace [Online]. Available at: https://arstechnica.com/tech-policy/2021/09/big-tech-companies-snap-up-smaller-rivals-at-record-pace/?comments=1.

Stahel, W.R. (1986). *The Functional Economy: Cultural and Organizational Change*. Science & Public Policy, London.

Staune, J. (2015). *Les clés du futur – Réinventer ensemble la société, l'économie et la science*. Plon, Paris.

Steenkamp, J.B.E.M. (2002). Global consumers. Presentation at Tilburg University. Based on consumer and market drivers of the trial probability of new consumer packaged goods. Working paper, University of Tilburg, Tilburg.

Stengers, I. (2014). Ralentir les sciences, c'est réveiller le chercheur somnambule. *Écologie & politique*, 48, 61–74.

Stengers, I. and Bensaude-Vincent, B. (2003). *100 mots pour commencer à penser les sciences*. Les empêcheurs de penser en rond, Paris.

Stiegler, B. (2014). À propos du verbe faire, des makers, des FabLab et de l'automatisation intégrale [Online]. Available at: https://www.centrale-marseille.fr/fr/news/une-ecole-ephemere-imaginee-philosophe-bernard-stiegler#sthash.WNm4geFQ.vIKrIOnh.dpuf.

Stiegler, B. (2021). *Nietzsche et la vie. Une nouvelle histoire de la philosophie*. Gallimard, Paris.

Sureau, F. (2022). Éléments du discours de F. Sureau à l'Académie Française rapportés par Richard F. Discours de François Sureau à l'Académie française : un éloge de la liberté [Online]. Available at: https://www.contrepoints.org/2022/03/08/422904-discours-de-francois-sureau-a-lacademie-francaise-un-eloge-de-la-liberte.

Szostak, B.L. (2021). Design thinking. Design thinking and strategic management of innovation. *Journal of Product Innovation Management* [Online]. Available at: https://onlinelibrary.wiley.com/doi/abs/10.1002/9781119832522.ch12.

Taguieff, P.A. (2021). *Les nietzschéens et leurs ennemis*. Édition du Cerf, Paris.

Taleb, N.N. (2013). *Anti-fragile*. Les Belles Lettres, Paris.

Tarde, G. (2001). *Les lois de l'imitation*. Le Seuil, Paris.

Taylor, F.W. (2006). Report of a lecture by and questions put to Mr. F.W. Taylor: A transcript. *Journal of Management History*, 14, 214–236.

Tenzer, N. (2011). *La fin du malheur français ? : Un nouveau devoir politique*. Stock, Paris.

Terrasse, P., Barbezieux, P., Herody, C. (2016). Rapport au premier ministre sur l'économie collaborative [Online]. Available at: http://www.gouvernement.fr/sites/default/files/liseuse/6421/master/projet/Rapport-de-Pascal-Terrasse-sur-le-développement-de-l-économie-collaborative.pdf.

Thibon, G. (2022). *Au secours des évidences*. MAME, Paris.

Townley, B. (2008). *Reason's Neglect: Rationality and Organizing*. Oxford University Press, Oxford.

Tripathi, N., Seppänen, P., Boominathan, G., Oivo, M., Liukkunen, K. (2018). Insights into start-up ecosystems through exploration of multi-vocal literature. *Information and Software Technology*, 105, 56–77.

Trott, P. (2008). *Innovation Management and New Product Development*. Prentice Hall, New York.

Tsui, A.S. (2007). From homogenization to pluralism: International management research in the academy and beyond. *Academy of Management Journal*, 50, 1353–1364.

Tsui, A.S. (2021). Responsible research and responsible leadership studies. *Academy of Management Discoveries*, 7, 166–170.

Tucsnak, M. and Weiss, G. (2014). Well-posed systems. The LTI case and beyond. *Automatica*, 50, 1757–1779.

USDC (2008). Innovation measurement, tracking the state of innovation in the American economy [Online]. Available at: https://www.commerce.gov/data-and-reports/reports/2008/01/innovation-measurement-tracking-state-innovation-american-economy.

Valéry, P. (1957). *Œuvres*. Gallimard, Paris.

Van de Ven, A.H., Polley, D., Garud, R., Venkatraman, S. (1999). *The Innovation Journey*. Oxford University Press, Oxford.

Van de Vliert, E. and Murray, D.R. (2018). Climate and creativity: Cold and heat trigger invention and innovation in richer populations. *Creativity Research Journal*, 30, 17–28.

Villani, E., Rasmussenr, E., Grimaldi, R. (2017). How intermediary organizations facilitate university–industry technology transfer: A proximity approach. *Technological Forecasting & Social Change*, 114, 86–102.

Vinatier, I. and Morrissette, J. (2015). Les recherches collaboratives : enjeux et perspectives. *Carrefours de l'éducation*, 1, 137–170.

Volkmann, C.K., Tokarsky, K.O., Grünhagen, M. (2010). *Entrepreneurship in a European Perspective: Concepts for the Creation and Gros of New Ventures*. Gabler Verlag, Berlin.

Vygotsky, L.S. (1999). *La signification historique de la crise en psychologie*. Delachaux et Niestlé, Paris.

Waikar, S. (2022). Think different. Sometimes. Teams succeed when they balance creativity and focus [Online]. Available at: https://www.gsb.stanford.edu/insights/think-different-sometimes-teams-succeed-when-they-balance-creativity-focus?utm_source=newsletter&utm_medium=email&utm_content=cognitive%20diversity&utm_campaign=Enews%20BOTW%201/28/2022.

Wannenmacher, D. (2020). Knowledge management et projets innovants. *Techniques de l'ingénieur*.

Weber, M. (2002). *Le savant et le politique*. 10/18, Paris.

Weick, K.E. (1995). *Sensemaking in Organizations*. Sage Publications, London.

Weil, S. (1955). *Oppression et liberté*. Gallimard, Paris.

Weil, B. (1999). *Conception collective, coordination et savoirs. Les rationalisations de la conception automobile*. École Nationale Supérieure des Mines de Paris, Paris.

Weinreich, S., Sahin, T., Inkermann, D., Huth, T., Vietor, T. (2020). Managing disruptive innovation by value-oriented portfolio planning. *International Design Conference – Design*, 1, 1395–1404.

Whitehead, A.N. (1975). *Science and the Modern World*. Cambridge University Press, Cambridge.

Whittaker, D.H., Sturgeon, T., Okita, T., Zhu, T. (eds) (2020). *Compressed Development: Time and Timing in Economic and Social Development*. Oxford University Press, Oxford.

Wiggins, G.A. (2006). Searching for computational creativity. *New Generation Computing*, 24, 209–222.

Wisner, A. (1995). *Réflexions sur l'ergonomie*. Octarès, Toulouse.

Wittrock, C., Forsberg, E.M., Pols, A., McNaghten, P., Ludwig, D. (2021). *Implementing Responsible Research and Innovation: Organizational and National Conditions*. Springer, Berlin.

Wright, M., Siegel, D.S., Mustar, P. (2017). An emerging ecosystem for student start-ups. *The Journal of Technology Transfer*, 42, 1–14.

Wu, L., Wang, D., Evans, J.A. (2019). Large teams develop and small teams disrupt science and technology. *Nature*, 566, 378–382.

Zerubavel, E. (1991). *The Fine Line. Making Distinctions in Everyday Life*. The University of Chicago Press, Chicago.

Zhou, C. and Wang, R. (2020). From invention to innovation: The role of knowledge-intensive business services in technology commercialization. *Technology Analysis & Strategic Management*, 32, 1436–1448.

Zubizarreta, M., Ganzarain, J., Cuadrado, J., Lizarralde, R. (2021). Evaluating disruptive innovation project management capabilities. *Sustainability*, 13, 1. doi: 10.3390/su13010001.

Conclusion of Volume 1

The management of research, the functioning of projects, the evaluation centered too much so on productions, the weakness of the management of interdisciplinary projects and the position of academic research vis-à-vis companies are questions raised in this first volume. At the same time, the somewhat hazardous and difficult link between idea and innovation, a key element in the development of emerging technologies full of promise, is discussed. This work, in a way prior to any link with society, is undoubtedly useful with its bottom-up logic and with a small number of actors to save time in producing a convincing proof of concept. But, obviously, this is not enough. And it is this aspect that will be discussed in Volume 2.

What was shown in this first volume is that there is no innovation without society, and certainly not against society because invention and innovation are intended for them. But, for all that, before linking the "happy" idea and the users, in real life, there are many obstacles to overcome which are often far from science, but some of which are part of it. To take an example, scientific research highlights issues of complexity in invention. It cannot stop there because the downstream sector, which is productive, needs clear procedures to achieve the objects, the processes resulting from this intellectual "fermentation".

On the one hand, there is a need to make radical transitions between complexity and determinism for production, and, on the other hand, attractive inventions are needed to make this production make industrial sense for interested customers. This means that, as soon as possible, these in-depth interactions between science open to society and industrial production open to science must be put in place, despite the various "Herculean tasks" presented in Volume 1. They must lead to a real mutualization of interests, perhaps one day leading to integration, a convergence of knowledge that is currently too disjointed.

Index

A, B

C

D

E, F

G, H, I

K, L, M

N, O, P

Q, R

S, T

U, V

Other titles from

ISTE

in

Systems and Industrial Engineering – Robotics

2023

BERRAH Lamia, CLIVILLÉ Vincent
Decision and Decision-maker in an Industrial Environment

BRIFFAUT Jean-Pierre
Complexities 1: Various Approaches in the Field of Techno-scientific Knowledge
(Systems of Systems Complexity – Volume 5)

CHAINEAUX Jacques
The Mechanisms of Explosions: 27 Case Studies for their Understanding

DIALLO Saikou, BOGUI Jean-Jacques Maomra, ATCHOUA Julien
Technologies and African Societies in Pandemic Times: Using Technology to Survive and Thrive During the Covid-19 Era
(Africa Digital Age Set – Volume 2)

JAEGLER Anicia, ROQUES Thierry
Sustainable Supply Chain Management: Sustainable Decision Support

PRINTZ Jacques
Organization and Pedagogy of Complexity: Systemic Case Studies and Prospects
(Systems of Systems Complexity Set – Volume 4)

2022

AMARA Yacine, BEN AHMED Hamid, GABSI Mohamed
Hybrid Excited Synchronous Machines: Topologies, Design and Analysis

BOURRIÈRES Jean-Paul, PINÈDE Nathalie, TRAORÉ Mamadou Kaba, ZACHAREWICZ Grégory
From Logistic Networks to Social Networks: Similarities, Specificities, Modeling, Evaluation

DEMOLY Frédéric, ANDRÉ Jean-Claude
4D Printing 1: Between Disruptive Research and Industrial Applications
4D Printing 2: Between Science and Technology

HAJJI Rafika, JARAR OULIDI Hassane
Building Information Modeling for a Smart and Sustainable Urban Space

KROB Daniel
Model-based Systems Architecting: Using CESAM to Architect Complex Systems
(Systems of Systems Complexity Set – Volume 3)

LOUIS Gilles
Dynamics of Aircraft Flight

2020

BRON Jean-Yves
System Requirements Engineering

KRYSINSKI TOMASZ, MALBURET FRANÇOIS
Energy and Motorization in the Automotive and Aeronautics Industries

PRINTZ Jacques
System Architecture and Complexity: Contribution of Systems of Systems to Systems Thinking

2019

ANDRÉ Jean-Claude
Industry 4.0: Paradoxes and Conflicts

BENSALAH Mounir, ELOUADI Abdelmajid, MHARZI Hassan
Railway Information Modeling RIM: The Track to Rail Modernization

BLUA Philippe, YALAOU Farouk, AMODEO Lionel, DE BLOCK Michaël, LAPLANCHE David
Hospital Logistics and e-Management: Digital Transition and Revolution

BRIFFAUT Jean-Pierre
From Complexity in the Natural Sciences to Complexity in Operations Management Systems
(Systems of Systems Complexity Set – Volume 1)

BUDINGER Marc, HAZYUK Ion, COÏC Clément
Multi-Physics Modeling of Technological Systems

FLAUS Jean-Marie
Cybersecurity of Industrial Systems

JAULIN Luc
Mobile Robotics – Second Edition Revised and Updated

KUMAR Kaushik, DAVIM Paulo J.
Optimization for Engineering Problems

TRIGEASSOU Jean-Claude, MAAMRI Nezha
Analysis, Modeling and Stability of Fractional Order Differential Systems 1: The Infinite State Approach
Analysis, Modeling and Stability of Fractional Order Differential Systems 2: The Infinite State Approach

VANDERHAEGEN Frédéric, MAAOUI Choubeila, SALLAK Mohamed, BERDJAG Denis
Automation Challenges of Socio-technical Systems

2018

BERRAH Lamia, CLIVILLÉ Vincent, FOULLOY Laurent
Industrial Objectives and Industrial Performance: Concepts and Fuzzy Handling

GONZALEZ-FELIU Jesus
Sustainable Urban Logistics: Planning and Evaluation

GROUS Ammar
Applied Mechanical Design

LEROY Alain
Production Availability and Reliability: Use in the Oil and Gas Industry

MARÉ Jean-Charles
Aerospace Actuators 3: European Commercial Aircraft and Tiltrotor Aircraft

MAXA Jean-Aimé, BEN MAHMOUD Mohamed Slim, LARRIEU Nicolas
Model-driven Development for Embedded Software: Application to Communications for Drone Swarm

MBIHI Jean
Analog Automation and Digital Feedback Control Techniques
Advanced Techniques and Technology of Computer-Aided Feedback Control

MORANA Joëlle
Logistics

SIMON Christophe, WEBER Philippe, SALLAK Mohamed
Data Uncertainty and Important Measures
(Systems Dependability Assessment Set – Volume 3)

TANIGUCHI Eiichi, THOMPSON Russell G.
City Logistics 1: New Opportunities and Challenges
City Logistics 2: Modeling and Planning Initiatives
City Logistics 3: Towards Sustainable and Liveable Cities

ZELM Martin, JAEKEL Frank-Walter, DOUMEINGTS Guy, WOLLSCHLAEGER Martin
Enterprise Interoperability: Smart Services and Business Impact of Enterprise Interoperability

2017

ANDRÉ Jean-Claude
From Additive Manufacturing to 3D/4D Printing 1: From Concepts to Achievements
From Additive Manufacturing to 3D/4D Printing 2: Current Techniques, Improvements and their Limitations
From Additive Manufacturing to 3D/4D Printing 3: Breakthrough Innovations: Programmable Material, 4D Printing and Bio-printing

ARCHIMÈDE Bernard, VALLESPIR Bruno
Enterprise Interoperability: INTEROP-PGSO Vision

CAMMAN Christelle, FIORE Claude, LIVOLSI Laurent, QUERRO Pascal
Supply Chain Management and Business Performance: The VASC Model

FEYEL Philippe
Robust Control, Optimization with Metaheuristics

MARÉ Jean-Charles
Aerospace Actuators 2: Signal-by-Wire and Power-by-Wire

POPESCU Dumitru, AMIRA Gharbi, STEFANOIU Dan, BORNE Pierre
Process Control Design for Industrial Applications

RÉVEILLAC Jean-Michel
Modeling and Simulation of Logistics Flows 1: Theory and Fundamentals
Modeling and Simulation of Logistics Flows 2: Dashboards, Traffic Planning and Management
Modeling and Simulation of Logistics Flows 3: Discrete and Continuous Flows in 2D/3D

2016

ANDRÉ Michel, SAMARAS Zissis
Energy and Environment
(Research for Innovative Transports Set – Volume 1)

AUBRY Jean-François, BRINZEI Nicolae, MAZOUNI Mohammed-Habib
Systems Dependability Assessment: Benefits of Petri Net Models
(Systems Dependability Assessment Set – Volume 1)

BLANQUART Corinne, CLAUSEN Uwe, JACOB Bernard
Towards Innovative Freight and Logistics
(Research for Innovative Transports Set – Volume 2)

COHEN Simon, YANNIS George
Traffic Management
(Research for Innovative Transports Set – Volume 3)

MARÉ Jean-Charles
Aerospace Actuators 1: Needs, Reliability and Hydraulic Power Solutions

REZG Nidhal, HAJEJ Zied, BOSCHIAN-CAMPANER Valerio
Production and Maintenance Optimization Problems: Logistic Constraints and Leasing Warranty Services

TORRENTI Jean-Michel, LA TORRE Francesca
Materials and Infrastructures 1
(Research for Innovative Transports Set – Volume 5A)
Materials and Infrastructures 2
(Research for Innovative Transports Set – Volume 5B)

WEBER Philippe, SIMON Christophe
Benefits of Bayesian Network Models
(Systems Dependability Assessment Set – Volume 2)

YANNIS George, COHEN Simon
Traffic Safety
(Research for Innovative Transports Set – Volume 4)

2015

AUBRY Jean-François, BRINZEI Nicolae
Systems Dependability Assessment: Modeling with Graphs and Finite State Automata

BOULANGER Jean-Louis
CENELEC 50128 and IEC 62279 Standards

BRIFFAUT Jean-Pierre
E-Enabled Operations Management

MISSIKOFF Michele, CANDUCCI Massimo, MAIDEN Neil
Enterprise Innovation

2014

CHETTO Maryline
Real-time Systems Scheduling
Volume 1 – Fundamentals
Volume 2 – Focuses

DAVIM J. Paulo
Machinability of Advanced Materials

ESTAMPE Dominique
Supply Chain Performance and Evaluation Models

FAVRE Bernard
Introduction to Sustainable Transports

GAUTHIER Michaël, ANDREFF Nicolas, DOMBRE Etienne
Intracorporeal Robotics: From Milliscale to Nanoscale

MICOUIN Patrice
Model Based Systems Engineering: Fundamentals and Methods

MILLOT Patrick
Designing Human–Machine Cooperation Systems

NI Zhenjiang, PACORET Céline, BENOSMAN Ryad, RÉGNIER Stéphane
Haptic Feedback Teleoperation of Optical Tweezers

OUSTALOUP Alain
Diversity and Non-integer Differentiation for System Dynamics

REZG Nidhal, DELLAGI Sofien, KHATAD Abdelhakim
Joint Optimization of Maintenance and Production Policies

STEFANOIU Dan, BORNE Pierre, POPESCU Dumitru, FILIP Florin Gh.,
EL KAMEL Abdelkader
Optimization in Engineering Sciences: Metaheuristics, Stochastic Methods and Decision Support

2013

ALAZARD Daniel
Reverse Engineering in Control Design

ARIOUI Hichem, NEHAOUA Lamri
Driving Simulation

CHADLI Mohammed, COPPIER Hervé
Command-control for Real-time Systems

DAAFOUZ Jamal, TARBOURIECH Sophie, SIGALOTTI Mario
Hybrid Systems with Constraints

FEYEL Philippe
Loop-shaping Robust Control

FLAUS Jean-Marie
Risk Analysis: Socio-technical and Industrial Systems

FRIBOURG Laurent, SOULAT Romain
Control of Switching Systems by Invariance Analysis: Application to Power Electronics

GROSSARD Mathieu, REGNIER Stéphane, CHAILLET Nicolas
Flexible Robotics: Applications to Multiscale Manipulations

GRUNN Emmanuel, PHAM Anh Tuan
Modeling of Complex Systems: Application to Aeronautical Dynamics

HABIB Maki K., DAVIM J. Paulo
Interdisciplinary Mechatronics: Engineering Science and Research Development

HAMMADI Slim, KSOURI Mekki
Multimodal Transport Systems

JARBOUI Bassem, SIARRY Patrick, TEGHEM Jacques
Metaheuristics for Production Scheduling

KIRILLOV Oleg N., PELINOVSKY Dmitry E.
Nonlinear Physical Systems

LE Vu Tuan Hieu, STOICA Cristina, ALAMO Teodoro,
CAMACHO Eduardo F., DUMUR Didier
Zonotopes: From Guaranteed State-estimation to Control

MACHADO Carolina, DAVIM J. Paulo
Management and Engineering Innovation

MORANA Joëlle
Sustainable Supply Chain Management

SANDOU Guillaume
Metaheuristic Optimization for the Design of Automatic Control Laws

STOICAN Florin, OLARU Sorin
Set-theoretic Fault Detection in Multisensor Systems

2012

AÏT-KADI Daoud, CHOUINARD Marc, MARCOTTE Suzanne, RIOPEL Diane
Sustainable Reverse Logistics Network: Engineering and Management

BORNE Pierre, POPESCU Dumitru, FILIP Florin G., STEFANOIU Dan
Optimization in Engineering Sciences: Exact Methods

CHADLI Mohammed, BORNE Pierre
Multiple Models Approach in Automation: Takagi-Sugeno Fuzzy Systems

DAVIM J. Paulo
Lasers in Manufacturing

DECLERCK Philippe
Discrete Event Systems in Dioid Algebra and Conventional Algebra

DOUMIATI Moustapha, CHARARA Ali, VICTORINO Alessandro, LECHNER Daniel
Vehicle Dynamics Estimation using Kalman Filtering: Experimental Validation

GUERRERO José A, LOZANO Rogelio
Flight Formation Control

HAMMADI Slim, KSOURI Mekki
Advanced Mobility and Transport Engineering

MAILLARD Pierre
Competitive Quality Strategies

MATTA Nada, VANDENBOOMGAERDE Yves, ARLAT Jean
Supervision and Safety of Complex Systems

POLER Raul *et al.*
Intelligent Non-hierarchical Manufacturing Networks

TROCCAZ Jocelyne
Medical Robotics

YALAOUI Alice, CHEHADE Hicham, YALAOUI Farouk, AMODEO Lionel
Optimization of Logistics

ZELM Martin *et al.*
Enterprise Interoperability –I-EASA12 Proceedings

2011

CANTOT Pascal, LUZEAUX Dominique
Simulation and Modeling of Systems of Systems

DAVIM J. Paulo
Mechatronics

DAVIM J. Paulo
Wood Machining

GROUS Ammar
Applied Metrology for Manufacturing Engineering

KOLSKI Christophe
Human–Computer Interactions in Transport

LUZEAUX Dominique, RUAULT Jean-René, WIPPLER Jean-Luc
Complex Systems and Systems of Systems Engineering

ZELM Martin, *et al.*
Enterprise Interoperability: IWEI2011 Proceedings

2010

BOTTA-GENOULAZ Valérie, CAMPAGNE Jean-Pierre, LLERENA Daniel, PELLEGRIN Claude
Supply Chain Performance: Collaboration, Alignment and Coordination

BOURLÈS Henri, GODFREY K.C. Kwan
Linear Systems

BOURRIÈRES Jean-Paul
Proceedings of CEISIE'09

CHAILLET Nicolas, REGNIER Stéphane
Microrobotics for Micromanipulation

DAVIM J. Paulo
Sustainable Manufacturing

GIORDANO Max, MATHIEU Luc, VILLENEUVE François
Product Life-Cycle Management: Geometric Variations

LOZANO Rogelio
Unmanned Aerial Vehicles: Embedded Control

LUZEAUX Dominique, RUAULT Jean-René
Systems of Systems

VILLENEUVE François, MATHIEU Luc
Geometric Tolerancing of Products

2009

DIAZ Michel
Petri Nets: Fundamental Models, Verification and Applications

OZEL Tugrul, DAVIM J. Paulo
Intelligent Machining

PITRAT Jacques
Artificial Beings

2008

ARTIGUES Christian, DEMASSEY Sophie, NERON Emmanuel
Resources–Constrained Project Scheduling

BILLAUT Jean-Charles, MOUKRIM Aziz, SANLAVILLE Eric
Flexibility and Robustness in Scheduling

DOCHAIN Denis
Bioprocess Control

LOPEZ Pierre, ROUBELLAT François
Production Scheduling

THIERRY Caroline, THOMAS André, BEL Gérard
Supply Chain Simulation and Management

2007

DE LARMINAT Philippe
Analysis and Control of Linear Systems

DOMBRE Etienne, KHALIL Wisama
Robot Manipulators

LAMNABHI Françoise *et al.*
Taming Heterogeneity and Complexity of Embedded Control

LIMNIOS Nikolaos
Fault Trees

2006

FRENCH COLLEGE OF METROLOGY
Metrology in Industry

NAJIM Kaddour
Control of Continuous Linear Systems

Printed and bound by CPI Group (UK) Ltd, Croydon, CR0 4YY
28/07/2026
14931340-0003